The Moral Foundation of Rights

L. W. SUMNER

CLARENDON PRESS · OXFORD

Oxford University Press, Walton Street, Oxford OX2 6DP
Oxford New York Toronto
Delhi Bombay Calcutta Madras Karachi
Petaling Jaya Singapore Hong Kong Tokyo
Nairobi Dar es Salaam Cape Town
Melbourne Auckland
and associated companies in
Berlin Ibadan

Oxford is a trade mark of Oxford University Press

Published in the United States
by Oxford University Press, New York

British Library Cataloguing in Publication Data
Sumner, L. W.
The moral foundation of rights.
1. Ethics
I. Title
170 BJ1012
ISBN 0–19–824751–6
ISBN 0–19–824874–1 (Pbk.)

Library of Congress Cataloging-in-Publication Data
Sumner, L. W.
The moral foundation of rights.
Bibliography: p.
Includes index.
1. Ethics I. Title. II. Title: Rights.
BJ1031.S86 1987 172 87–5640
ISBN 0–19–824751–6
ISBN 0–19–824874–1 (Pbk.)

Printed and bound in Great Britain by
Biddles Ltd
Guildford and King's Lynn

FOR HEATHER

PREFACE

THE argument of this book begins with a moral concept and ends with a moral theory; its central theme is the relationship between the two. Neither the choice of concept nor the choice of theory was arbitrary. Rights are on their way to becoming the accepted international currency of moral, and especially political, debate. They have also attracted a great deal of attention (favourable and unfavourable) in the recent philosophical literature, especially in the English-speaking world. In liberal societies, and perhaps elsewhere as well, both the philosophical community and the public have come to take rights very seriously indeed. Nor is this love affair likely to prove a mere passing infatuation. On the one hand the concept of a right is politically attractive because it seems peculiarly well suited to expressing some of the ideals at the heart of liberal political theories. As long as such theories remain in fashion rights will also remain in fashion. On the other hand the notion of a moral right is sufficiently intricate and puzzling to provide a challenge to philosophers, both those who are friendly to the liberal tradition and those who are not. Thus both in politics and in philosophy rights seem to be here to stay.

Despite some premature notices of their demise, consequentialist moral theories also seem to be here to stay. But one of the obstacles to their acceptance has always been their apparent hostility to such paradigmatically deontological values as rights. What distinguishes consequentialist theories from their rivals is that they are goal-based—that is, at bottom they counsel the pursuit of some global, synoptic goal. By contrast, rights appear to function normatively as constraints on the pursuit of such goals. The main theoretical conclusion to be defended in what follows is that the supposed incompatibility between commitment to a basic goal and acceptance of constraints on the pursuit of that goal is an illusion. Any viable form of consequentialism, when combined with a realistic picture of the nature of moral agents and of the world within which they operate, must make room for rights. The appearance to the contrary has been fostered chiefly by an oversimplified view of the structure of consequentialism.

The thesis that a consequentialist theory can protect the integrity of rights is controversial. The thesis that no non-consequentialist theory can do so is even more controversial. Although I believe that this companion thesis is also true, I will not undertake a full defence of it here. There are many forms of non-consequentialism. A full defence of the second thesis would require showing that every such theory is incapable of supporting rights. That enterprise is, for obvious reasons, beyond the resources of this book. I will therefore content myself with a partial defence which will consist of showing that two prominent varieties of non-consequentialism—those which are commonly thought to be friendliest to rights—cannot do the job. If they fail then we have better reason to think that only a consequentialist theory can succeed. But that reason will not be conclusive.

An independent case can be made both for the importance of rights and for the plausibility of consequentialism. The affiliation between the concept and the theory, however, strengthens the case for each. On the one hand, consequentialism is capable of revealing the point or rationale of rights. On the other hand, finding room for rights within a consequentialist framework helps to dispel the impression that consequentialism requires the abandonment of some of our most deeply entrenched moral/political values. If the argument of this book is successful a practical commitment to the protection of rights is compatible with, indeed required by, a theoretical commitment to the pursuit of goals.

There is another, more biographical, respect in which my selection of concept and theory is not arbitrary. This book owes its origin in part to some unfinished philosophical business. In *Abortion and Moral Theory* I defended a moderate view of the morality of abortion, in part by trying to show that it coheres well with a utilitarian moral framework. This programme was imperilled, however, by the fact that views of abortion are standardly couched in terms of rights, both those of women and those of fetuses. Since rights appear to be alien to utilitarianism, I faced the problem of showing that the moral theory I espouse was capable of supporting any position on abortion, moderate or otherwise. One alternative open to me was to reject all talk of rights as so much moral superstition and recast my view of abortion in some other terms.

But I preferred another path which involved grounding conclusions about rights, and especially about the right to life of fetuses, on a utilitarian foundation. Accordingly, I set about outlining a version of utilitarianism which seemed well adapted to this task and then tried to show that it would support my moderate views about fetal rights, and thus also about the morality of abortion. I am not very happy with the result. Undertaking a theoretical enterprise of this scope in the course of a book which had a very narrow practical focus seems to me in retrospect to have been quixotic. Thus I have resolved to tackle the theoretical issues in their own right, largely in abstraction from their practical implications. This book is the result.

The project of trying to reconcile rights and consequentialism was also motivated by the fact I am committed to both. In my everyday moral life I believe in various sorts of moral rights and support organizations which aim to promote respect for those rights. Meanwhile, however, in my philosophical life I profess a theory which has a reputation for hostility to moral rights of any kind. The present undertaking is therefore also an attempt to unite the two sides of my apparently divided moral self.

Whatever its origins, my project could not have been carried out without the assistance of a variety of organizations and individuals. The first draft was written during two years in Oxford, from 1981 to 1983, which were made possible by a Killam Research Fellowship from the Canada Council and a Research Grant from the Social Sciences and Humanities Research Council of Canada. I am very grateful to both of these agencies for their financial support, and to the Subfaculty of Philosophy at Oxford University for the amenities which it made available to me. A considerable portion of the final draft was written during the spring semester of 1986, which I spent as Distinguished Visiting Professor at Bowling Green State University. I wish to record my gratitude to the Bowling Green Philosophy Department for their characteristic hospitality during my term there, and to my home department at the University of Toronto for its patience in allowing me to take up these various opportunities. I am also grateful to the University of Toronto Faculty of Law for generously offering me a research grant under the auspices of their Connaught Programme in Legal Theory and Public Policy.

Over the past few years I have presented a number of papers

containing ideas which have been worked into this book. I know that I have learned a great deal from the members of those audiences, but I fear that I can no longer disentangle their contributions and thus can thank them only collectively. Other debts, however, can be more readily individuated. Various chapters of the first draft elicited valuable responses from Jerry Bickenbach, Ronald Dworkin, R. G. Frey, Joel Kupperman, Christopher Morris, Joseph Raz, and E. J. Weinrib. I owe a special debt to Carl Wellman, whose detailed comments on the entire draft made the task of rewriting it both harder and easier. In addition, during my stay in Oxford I profited from many discussions with James Griffin, whose views helped me to clarify my own both when we agreed and when we disagreed. Finally, Tom Hurka and Joseph Raz provided very helpful criticisms of a portion of the final draft. I have no doubt that had I shown them more of it they would have shown me more of its mistakes.

Most of the material in this book appears in print for the first time. However, chapter four revises and expands my essay 'Rights Denaturalized', which was published in *Utility and Rights*, edited by R. G. Frey. I am grateful to the University of Minnesota Press for permission to use this material. In addition, section 6.2 of chapter six incorporates a considerably rearranged version of 'Utilitarian Goals and Kantian Constraints', which is to appear in B. Brody, ed., *Moral Theory and Moral Judgments* (Copyright © 1988 by D. Reidel Publishing Company, Dordrecht, Holland).

Finally, I wish to record my gratitude to my wife, Heather Wright, for her patience during the more difficult stretches of the production process. In dedicating the eventual result to her I also undertake to mend my ways.

L.W.S.
Toronto
August 1986

CONTENTS

I

The Problem of Rights

LIKE the arms race the escalation of rights rhetoric is out of control. In the liberal democracies of the West, and especially in the United States, public issues are now routinely phrased in the language of rights. One of the most visible examples of this proliferation, and also one of its earliest casualties, has been the contemporary debate about the morality of abortion in which the 'pro-choice' appeal to women's right to control their own reproduction is countered with tedious predictability by the 'pro-life' appeal to the fetus's right to life. But there is virtually no area of public controversy in which rights are not to be found on at least one side of the question—and generally on both.

In education, for example, compulsory schooling to some stipulated minimum age is commonly defended as respecting the right of children to be offered a curriculum which will adequately equip them for adult life and just as commonly attacked as denying the right of parents to control their children's upbringing. This contest between the rights of children and their parents surfaces as well in debates about the funding of a state school system and about control of the curriculum of that system. Meanwhile, the right of parents to choose their children's schools also faces competition from the right of members of racial or religious minorities not to be disadvantaged by second-class schooling. The busing programmes which are now familiar features of the American educational system limit the right of parents to send their children to neighbourhood schools (or the right of the children to attend such schools) in an attempt to ensure an equitable distribution of educational resources.

In the area of health care we encounter similar conflicts of similar rights. The child/parent dichotomy is reproduced in controversies over the medical treatment of minors, most acutely in cases in which (for whatever reason) parents decline consent to life-saving therapy for their children but also in cases in which a physician protects the confidentiality of an underage girl in prescribing a contraceptive to which the girl's parents are known to object. Likewise, the rights of patients may collide with the rights of their guardians when mentally retarded girls or women are subjected to sterilizations to which they are incompetent to consent. But the more general conflict of rights in the health care system is between consumers and providers. On the level of a particular patient/physician relationship this conflict can pit the patient's right to decline treatment (even where doing so may threaten life) against the physician's duty to provide standard care or not to harm the patient's interests. Where therapy becomes experimental the right of a subject to give free and informed consent may collide with the necessity of deception as an integral part of a scientifically valid research protocol. On the broader social level the rights of health care practitioners may be curtailed in various ways in order to ensure an equitable distribution of their services. Thus, for example, physicians may be denied the opportunity to charge what the market will bear for their services by the imposition of a public medical insurance plan with a compulsory fee schedule. Such schemes, which are standardly attacked as denials of freedom of contract between physicians and patients, are just as standardly defended as guarantees of everyone's right to at least a baseline standard of health care, regardless of the ability to pay for it.

Employment is an equally fertile ground for appeals to rights. On the side of labour the institutions of the closed shop and the picket line are defended on the ground that they ensure the democratic right of the majority to impose its will on the minority and attacked on the ground that they violate the individual's freedom of association (or dissociation). In the determination of employment conditions the right of management to run an efficient enterprise may compromise the right of employees to safe working conditions, or indeed to retention of their jobs. Likewise, the imposition of fair wage standards, such as equal pay for work of equal value, limits the right of management to offer what the market will bear but does so in order to protect vulnerable groups such as women from discriminatory treatment.

We could go on, sector by sector. In the housing market the rights of landlords collide with those of tenants, the rights of families with those of childless couples, and the rights of developers with those of homeowners. In the administration of justice the rights of the police clash with those of private citizens, the rights of offenders with those of victims, and the rights of prisoners with those of prison officials. In environmental controversies the rights of pollutors are opposed by the rights of those who are downstream or downwind from their pollution, the rights of extractors by those of other users of wilderness areas, and rights of humans by those of animals. Indeed, liberal societies appear to be replete with conflicts of rights: the young against the old, one race against another, natives against foreigners, the rich against the poor, men against women, humans against animals, one religious sect against another, believers against atheists, smokers against non-smokers, parents against children, the present generation against future generations, gays against straights, individuals against collectivities, one linguistic group against another, the media against the government, citizens against the police, employers against employees, opera-lovers against baseball fans, the public sector against the private, country-dwellers against city-dwellers, motorists against pedestrians, producers against consumers, white-collar workers against blue-collar, puritans against libertines, families against the childless, the healthy against the handicapped, seniors against juniors, agriculture against industry, jobholders against the jobless, teachers against students, and everyone against the state.

In the international arena, meanwhile, the language of human rights has become the accepted currency for both the criticism and defence of the economic, diplomatic, cultural, humanitarian, and military policies of nation states. As in the case of domestic politics, rights are routinely invoked on both sides of most international controversies. Thus the right of national sovereignty is played off against the right to intervene on behalf of oppressed minorities, the right to protect domestic industries is countered by the right of free trade, the right of poor nations to development aid is balanced against the right of rich nations to determine the disposition of their own national product, the right of nations to a return on their educational investment is invoked against the right of individual citizens to emigrate, the right of self-defence is appealed to in order to justify military adventures abroad, and in general the rights of the strong conflict with those of the weak, the rights of the developed with those of the underdeveloped, the rights of aggressors with those of defenders, maritime nations with the land-locked, the resource-rich with the resource-poor, exporters with importers, industrial nations with agricultural, the north with the south, creditors with debtors, nuclear nations with non-nuclear, the colonial with the imperial, the culturally dominant with the dominated, and all nations with supranational organizations.

If we once pause to reflect on the bewildering array of rights invoked in both domestic and international affairs we cannot avoid asking ourselves some hard questions. Which of these rights are genuine and which are not? Which deserve to be taken seriously and which do not? And in cases in which there is a genuine right on each side of an issue, which deserves to be taken more seriously? In order to answer these questions we need some resource which will enable us both to assess and to arbitrate among rights claims. No simple appeal to intuitions seems likely to suffice for these purposes, since all rights claims are presumably found intuitively appealing at least by those who advance them. In recent history one attempt to provide the needed resource has taken the form of issuing a manifesto or catalogue of basic human rights. Many individual states have of course built some such catalogue into their legal system, in the form of a charter or bill of rights.[1] But the more novel and dramatic development since the Second World War has been the appearance of an array of declarations, convenants, conventions, and charters which can now be said to constitute an international legal code of human rights.[2] The purpose of all these general instruments has been to establish some common, ideally global, standards of just or decent conduct. They are impressive documents, in some cases inspiring ones, and their existence has done much to mitigate cruel, degrading, and oppressive practices. The advantage of a catalogue is that the rights it contains are more general than those standardly invoked in particular public issues, thus raising the promise of verifying narrower claims by subsuming them under some broader category. Presumably we would then be entitled to reject as spurious any putative specific right which could not be derived from some abstract and general right. A rights catalogue would thus enable us to ascertain the authenticity of rights in somewhat the way in which a stamp catalogue does so for stamps. It would serve as an authority.

[1] For an encapsulated summary of the protection of civil and political rights by individual states see Humana 1984.

[2] The principal ingredients of this code can be found in Brownlie 1981 and Sieghart 1983. For an accessible overview see Sieghart 1985.

However, several features of rights catalogues prevent them from actually functioning in this way. For one thing, as the rights which they list become more general they also become more vague, thereby compromising their power to confirm or disconfirm more specific claims. The Universal Declaration of Human Rights, probably the best known international catalogue, will serve to illustrate this problem. Article 3 states that 'Everyone has the right to life, liberty and security of person', Article 6 tells us that 'Everyone has the right to recognition everywhere as a person before the law', and Article 7 adds the provision that 'All are equal before the law and are entitled without any discrimination to equal protection of the law'. What is the scope here of 'everyone' and 'all'? Since these articles occur in a declaration of *human* rights, presumably we are to understand the quantifiers as applying only to human beings. Does this then imply that non-human animals have no rights worthy of legal protection? And do the quantifiers then include all human beings 'without any discrimination'? These provisions are presumably intended to prohibit discrimination on grounds, say, of race or gender or religious affiliation. But do they prohibit discrimination on grounds of age? Do human fetuses count here as human beings? Are they included within the scope of legal personhood, or the equal protection of life? If so, then the Declaration will authenticate the right to life of the fetus which is urged by anti-abortion groups; if not, then it will not. As it stands, however, it is simply silent on this question. It can therefore be used neither to confirm nor to disconfirm appeals to fetal rights. Furthermore, it will be disabled in this way whenever the rights which it contains are vague or indeterminate. Thus while reference to the Declaration might authenticate some specific rights in relatively easy cases, it will fail us in precisely the cases for which we most need a standard of authenticity.

We should expect indeterminacy of this sort in the case of any rights catalogue which lacks an institution empowered to render authoritative interpretations of its provisions. The rights enshrined in the municipal legal systems of nation states are usually as vague in their letter as those of the Universal Declaration. The needed interpretation, however, is provided by courts or other official bodies which have the authority to apply abstract rights to concrete cases. Once a sufficient density of adjudication has been built up, the rights as interpreted and applied in practice may be quite determinate. Thus in a particular jurisdiction it may be quite clear that fetuses do not count as persons before the law and thus lack the protection of life guaranteed to such persons. But this remedy for the problem of vagueness, which is possible only when a rights catalogue is legally binding in some jurisdiction, merely creates a compensating defect. Where a charter of rights is binding within a jurisdiction then it will serve to define the legal rights of those within that jurisdiction. But legal rights are the creatures of legal institutions, and especially of legislatures and courts. Since legal institutions and legal systems are themselves subject to assessment for the extent to which they respect basic rights, their pronouncements cannot provide an authoritative standard of authenticity for such rights. In normative debates on public issues appeals to rights are not, or are not only, appeals to legal rights. Instead, or additionally, they are appeals to moral rights which cannot, by their nature, be authoritatively declared by any legal institution.

Thus a catalogue of rights seems condemned to being either indeterminate or merely conventional, the former if it is not legally binding and the latter if it is. The main international instruments of human rights seem to have barely begun the long journey from the one status to the other.[3] To the extent that some authoritative body has been empowered to interpret the abstract rights which they contain, and apply them to participating states, those rights will gradually become determinate but also merely conventional. On the other hand, to the extent that individual states remain free to interpret these rights as they see fit, then the rights will remain indeterminate, indeed essentially contested. In fact an appropriate degree of indeterminacy seems often a precondition of securing agreement to the instruments by states with divergent political systems and political traditions. By way of example, the two United Nations Covenants govern, respectively, civil and political rights and economic, social, and cultural rights. It is well known that this distinction reflects the division of the post-war world between the Western powers who cherish the former and the Eastern bloc who prefer the latter. In effect, each side has agreed to recognize the other side's favourite rights in return for recognition of its own. The tacit presupposition of this log-rolling is that the two sides agree to disagree about the interpretation of both sets of rights. Thus the

[3] See Sieghart 1985, especially ch. 10.

Western powers claim that guaranteeing the right to work is compatible with soaring unemployment rates while the Eastern bloc maintains that protecting freedom of expression is compatible with making it a criminal offence to slander the state. Clearly any rights catalogue whose contents are open to this kind of manipulation cannot provide a resource capable of verifying specific rights claims.

Some rights manifestos are issued by bodies, such as private human rights groups, which make no pretence of being able to enforce them. As such, they may escape the dilemma confronting the more official instruments. Lacking any institutional status, the rights contained in these documents are plainly intended to formulate moral standards and are thus the sorts of abstract rights which might serve to authenticate more concrete ones. At the same time they are in principle capable of interpretation, and of application to particular cases, by the issuing bodies. Thus, for example, when Amnesty International declares that everyone has the right not to be imprisoned for reasons of conscience it must decide what is to count as a reason of conscience. Such a right need not therefore be vague. However, the basic defect which afflicts the more official catalogues also afflicts the less official ones. Appeals to the former cannot settle disputes about moral rights because these catalogues merely codify the outcomes of political agreements or the decisions of particular institutions, and the existence of moral rights cannot be established either by such agreements or by such decisions. Likewise, appeals to the latter cannot settle disputes about moral rights because these catalogues merely codify the commitments of one particular group, and the existence of moral rights also cannot be established by any such commitments. Although international agreements, authoritative legal decisions, and conscientious convictions may all count as evidence in favour of the existence of a moral right, this evidence cannot in the nature of the case be conclusive. Agreements, decisions, and convictions may all be morally justified or unjustified, and a confirmation procedure for rights claims must enable us to distinguish the former from the latter. No simple appeal to any catalogue of rights, legal or non-legal, can provide such a procedure.

We should also avoid overstating the extent to which unofficial rights catalogues agree in their contents. The reality is that rival manifestos simply replicate at a higher level of abstraction the conflicts among more specific rights claims. Do fetuses have a right to life? Some manifestos say yes and others say no. Do animals have rights? Do human beings have basic subsistence rights? Is there a right to private ownership of productive facilities? Again it is a simple matter to locate catalogues on both sides of each of these questions. In the end the deficiency shared by all catalogues is that they make no case in support of their contents. Rights claims must be confirmed or disconfirmed by argument and manifestos contain declarations in place of arguments. The shift from the arena of specific claims to the arena of general catalogues brings us no closer to a resource capable of distinguishing between justified and unjustified claims. Any such resources must be more than merely general and abstract; it must also provide, and justify, an argumentative framework.

The need for some such framework is now acute. It is the agility of rights, their talent for turning up on both sides of an issue, which is simultaneously their most impressive and their most troubling feature. Clearly interest groups which converge on little else agree that rights are indispensable weapons in political debate. This agreement reflects a deeper consensus concerning the peremptory nature of rights. As many commentators have noted, to say that I have a right to some good or service is not to say that it would be nice or generous or noble of others to give it to me; it is rather to say that they are obliged to do so, that it would be unfair or unjust of them not to, that I am entitled to expect or demand it of them, and so on. Thus the appeal to a right on one side of an issue must be countered by some equally potent weapon on the other. If instead it is opposed only by some less insistent appeal—to a mere good or ideal, for example—it will tend to prevail. But then if one interest group has built its case on an alleged right none of its competitors can.afford not to follow suit. Like any other weapons, once they have appeared in the public arena rights claims will tend both to proliferate and to escalate. The current inflation of rights rhetoric is thus an inevitable by-product of the pluralist political market-place.

In an arms race it can be better for each side to increase its stock though the resulting escalation will make all sides worse off. Where military weapons are concerned the increased threat is that of mutual annihilation. Where rhetorical weapons are concerned what all sides must fear is a backlash of scepticism or cynicism. An argumentative device capable of justifying anything is capable of justifying nothing. When rights claims have once been deployed on all sides of all public issues then they may no longer be taken seriously as means of resolving those issues. Indeed, the danger is that they will no longer be taken seriously at all. Just as fiscal inflation reduces the real value of money, the inflation of rights rhetoric threatens to erode the argumentative power of rights.

The ultimate form of cynicism about rights is nihilism. The nihilist hath said in his heart 'There are no rights'. (Of course, nihilists may also say 'There are no other values either', but we are concerned here with reasons for being especially sceptical about rights; more sweeping forms of moral nihilism therefore need not detain us.) Rights nihilism may have many sources and take many forms: Marxists may find rights too bourgeois, conservatives may find them too liberal, communitarians may find them too individual-istic, Europeans may find them too American, and consequentialists may find them too deontological. Whatever the nature of the complaint, its effect is to deny either the existence of rights or their importance in moral/political debate. At this stage of our inquiry we cannot dismiss these sceptical doubts out of hand. For all we now know, the very concept of a moral right may make no sense. Or perhaps it does make sense but there simply are no such things. Or perhaps there are such things but they deserve only a peripheral role in moral/political argument. Nihilism might, for all we now know, be the destination of our inquiry. Nihilism about any sort of thing is the challenge to make a good case for the existence of such things. One of my aims in this inquiry is to make a good case for the existence of rights. If in the end I have been unable to do so, despite my best efforts, then there will be little alternative to falling back on the nihilist view. Rights nihilism is thus the default position of this investigation. But just as nihilism cannot simply be assumed at the outset to be mistaken, neither can it simply be assumed to be correct. Indeed, the proliferation of rights claims which threatens to debase the coinage also tells us just how deeply embedded rights are in our current moral/political thinking. For most of us most of the time it is virtually unthinkable that there should be no rights or that the rights we take for granted should have only slight weight in the moral scales. Thinking does not of course make it so, and for all their entrenchment in our everyday practices it remains possible that rights are a mere fiction or illusion. But the intuitive evidence, for what it is worth, lends scant support to the nihilist view. Thus while nihilism is a possible end-point of the inquiry it is not an appropriate starting-point.

The possibility, indeed the reality, of nihilism about rights lends some urgency to the inquiry. It also defines the tasks at hand. If the integrity of rights is to be rescued from the nihilist's challenge then we must be able to impose some effective controls on the proliferation of rights claims. These controls must take the form of a verification procedure for such claims, or a standard of authenticity for rights. However, the notion of such a procedure or standard remains vague and undeveloped. It can be sharpened somewhat by imagining the various kinds of sceptical challenge which might be directed against a specific rights claim. Suppose I claim that citizens of poor nations have a moral right to development aid from citizens of affluent nations. The nihilist's challenges, which generalize across all rights, may be divided into two categories. The first challenge takes the form of contending that the concept of a moral right is incoherent or nonsensical. Members of poor nations cannot have a right to assistance because it makes no sense to say that anyone has any rights. A response to this challenge clearly must defend the intelligibility of at least some rights claims and the only convincing way to accomplish this is to construct a coherent concept of a moral right. Such a concept is one indispensable component of a standard of authenticity for rights. A putative right will fail to meet this purely conceptual standard if the very idea of it makes no sense. Thus even if we show that some rights claims make sense it will still be possible for a particular claim to make no sense.

Meaningfulness is necessary for truthfulness but it is not sufficient. A particular rights claim, or all such claims, may be coherent but false. This then brings us to the nihilist's second general challenge, which is not conceptual but substantive. Members of poor nations do not have a right to assistance because, as it happens, no one has any rights. A response to this challenge clearly must defend the truth of at least some rights claims and the only convincing way to accomplish this is to provide truth conditions for claims about the existence of rights—or, as I shall say, existence conditions for rights. The existence conditions for a given right tell us what must be the case in order for that right to exist. Such conditions will constitute the second ingredient in a standard of authenticity for rights. A putative right will fail to meet this substantive standard if its existence conditions fail to obtain. Thus even if we can show that some rights claims are true it will still be possible for a particular claim to be false.

A standard of authenticity must therefore include both a conceptual and a substantive component, the former telling us what rights are and the latter telling us which rights there are. In practice, however, it will be sufficient to develop existence conditions for rights since any adequate set of such conditions must incorporate some coherent concept of a right. If the very idea of a right is incoherent then rights can have no existence conditions. Furthermore, developing such a concept seems an indispensable step toward developing existence conditions since the concept will at least partially stipulate what must be the case in the world if it is to apply. A set of existence conditions for rights will therefore provide a standard of authenticity for rights and a verification procedure for rights claims. The search for a decisive rebuttal to the nihilist's challenges is the search for such conditions.

Not all challenges to rights claims are, however, as broad and sweeping as those of the nihilist. It is quite possible to concede both the intelligibility and the actual existence of some rights while challenging a particular instance, such as the right to development aid. We can sharpen the notion of existence conditions for rights still further by identifying the various forms such a challenge might take. These forms depend on the various dimensions of a right. The principal division here is between a right's *content* and its *scope*. The content of a right consists of whatever it is a right *to*. The content of the right I have claimed is to be given development aid. To challenge the content of my alleged right would be to deny that anyone has a right with that content. The scope of a right consists of the class of things whose normative positions are stipulated by the right. This class decomposes in turn into two subclasses. The *subjects* of a right are those who hold it. The subjects of the right I have claimed are members of poor nations. To challenge the subjects of my alleged right would be to deny that these people hold a right with that content. The *objects* of a right are those against whom it is held. The objects of the right I have claimed are members of affluent nations. To challenge the objects of my alleged right would be to deny that a right with that content holds against these people. My alleged right can therefore be faulted on the ground of its content, its subjects, or its objects (or any combination of these). Furthermore, each of these criticisms is in itself quite limited. To deny that anyone has a right *to development aid* is not to deny that people have other rights. To deny that *members of poor nations* have a right to development aid is not to deny that others have that right. And to deny that the right to development aid holds against *members of affluent nations* is not to deny that it holds against others. These specific challenges to my rights claim are therefore much more modest than the nihilist's general challenge, which is equivalent to reducing both the content and scope of every right to zero.

If a set of existence conditions is to serve as a standard of authenticity for rights then it must be capable, at least in principle, of testing both the content and the scope of every right alleged in every rights claim, thus telling us which rights anyone has, who has them, and against whom they hold. Nor is this all. Rights claims can be challenged on the ground not only of their truth but also of their importance. Thus it could be conceded that members of poor nations do indeed have the right to development aid but urged that this right has only slight weight, or even none at all, against competing considerations. Rights therefore have a fourth dimension as well, namely their *strength*. Clearly a right is not worth taking seriously if it is incapable of resisting many, or any, rival considerations (whether rights or other factors). The nihilist's general challenge is therefore also equivalent to reducing the weight of every right to zero.

We have already noted that rights function normatively as relatively insistent or peremptory moral considerations. Thus assigning only slight weight to a right will often be tantamount to denying the existence of the right. If the right to development aid can be overridden by virtually any competing consideration (property rights, the interests of affluent nations, the interests of affluent individuals, or whatever) then this is tantamount to denying that there is any such right. Specifying the strength of a right is important, however, in those cases in which a right competes with another right, or with some other similarly urgent consideration. As we have seen, rights are notorious for their ability to turn up on both sides of most moral/political issues. While many of these alleged rights will presumably be disqualified by a standard of authenticity, it seems unlikely that in each case only one of the competing rights will be genuine. However, if there are cases in which genuine rights conflict then we need to know which of the competitors has priority over the others (and to what extent).

A complete set of existence conditions for rights would be capable of specifying the content, scope, and strength of every authentic right, as well as locating the defect in every inauthentic right. Perhaps it is possible in principle to develop such conditions. If it could be done then the integrity of rights would be fully secure against both the spiralling inflation of their own rhetoric and the resultant nihilist backlash. For practical purposes, however, we may not need a confirmation procedure capable of yielding a complete answer for every case. In order to rebut the nihilist it is necessary only to provide a complete answer for some cases, or a partial answer for all cases, or even a partial answer for some cases. The more powerful the procedure we can develop the more secure rights will be as elements of moral/political argument. But we must not ask too much: there are limits to the degree of determinacy which it is reasonable to expect in morality and politics.

Thus far we have only questions and no answers. But our earlier discussion of rights catalogues has served to highlight one of the most puzzling features of moral rights—a feature which any set of existence conditions must somehow accommodate and explain. This feature can best be illustrated by contrasting moral and legal rights. On the one hand, we speak of both sorts of rights as things which can be possessed, exercised, demanded, waived, relinquished, transferred, and so on. Thus just as we say that black Americans have the legal right to full representation in the political system which governs them, so we say that black South Africans have the moral right to such representation. The grammar of both claims appears to be identical. If we think that legal rights are things whose possession advantages their owners legally then we will want to say that moral rights are things whose possession advantages their owners morally. Thus on the one hand a moral right appears to be a certain sort of commodity. However, it is clear that moral rights are not simply legal rights, or conventional rights of any kind. We say that black South Africans have the moral right to full representation even though this right has not been accorded legal recognition, and in saying this we mean to point to the right as a moral reason for changing the legal system so as to accord it recognition. Thus on the other hand moral rights provide moral reasons for establishing or maintaining conventional rights; they have, as I shall say, *moral force*.[4] The question which a set of existence conditions for moral rights must somehow answer is how a commodity palpable enough to be manipulated in various ways can have moral force. To the extent that moral rights are commodities, how do they differ from conventional rights? And to the extent that moral rights have moral force, how can they be commodities of any kind?

---

[4] I borrow this notion from Lyons 1982.

2

The Analysis of Rights

INFLATION devalues a currency by eroding its purchasing power. The proliferation of rights claims has devalued rights by eroding their argumentative power. The principal factor in this proliferation— the advancing of rights claims which, while meaningful, are groundless or frivolous—can be controlled only by a full set of existence conditions for rights. But a special contribution has been made by the distortion or abuse of the very language of rights. We can begin to control this factor by constructing an adequate concept of a moral right.

A complex moral concept can doubtless be abused in many different ways. The ambiguity of such concepts as a person or an interest, for example, furnishes boundless opportunities for equivocation. While the concept of a right is scarcely free of ambiguities, the abuse to which it lends itself is rather different. The normative function of the language of rights is to formulate one kind of urgent or insistent demand. The very urgency of rights, however, ensures that they will be appropriated by all sides on every issue and thus that they will come in time to serve as the common currency of moral/political argument. Thus while the concept of a right may begin its life as a specialized instrument the pressures of the political market-place will inexorably broaden it into a general-purpose notion. As a concept is stretched further and further beyond its proper domain it is also emptied of more and more of its distinctive content. Thus the increasing versatility of rights has been purchased at the cost of their increasing vacuity.

The contemporary philosophical literature yields some striking examples of this broadened, and correspondingly impoverished, notion of a right. Philosophers on both sides of the abortion debate, for example, tend to agree that the central issue in the debate is the moral status of the fetus. That issue is then commonly formulated in terms of whether the fetus is a person, and the notion of a person is in turn commonly defined in terms of the possession of rights.[1] By these successive stages the question whether the fetus counts for something or matters morally—whether it has moral standing—is simply identified with the question whether it has rights. The same reflex identification occurs in many environmental discussions in which the central issue once again is the distribution of moral standing. When philosophers ask whether fetuses or animals or organisms or biotic communities have rights, and when they do not treat this question as reflecting their adoption of one particular conceptual framework, then they are unthinkingly assuming that the language of rights is the language of morals.

But the best illustration of this conceptual expansion is provided by contemporary catalogues of human rights. The purpose of such catalogues is generally to formulate some common moral standards for assessing the policies of particular nation states. Their contents typically divide into two more or less discrete categories.[2] On the one hand civil and political rights protect values such as basic liberties, due process of law, and participation in the political system. However these rights may be specified or interpreted, their inclusion is clearly intended to impose upon governments the duty to ensure their full enjoyment by each individual citizen considered separately. Their denial to anyone, special circumstances aside, will therefore count as an injustice. It is these conceptual connections between the notion of a right and the cognate notions of duty and justice which give rights their distinctive cutting edge. By contrast, however, the catalogues also typically contain an array of economic, social, and cultural rights covering such matters as the necessaries of life, employment, social security, health care, and education. In the case of these rights we would normally be more reluctant to say that governments are obligated to ensure their full enjoyment by all citizens, or that the failure to do so in any particular case would necessarily constitute an injustice.[3] Our reluctance to say these things stems in part from our recognition that ensuring everyone's full enjoyment of these rights is likely to be beyond the resources of any but the wealthiest nations. Because governments can have a legitimate excuse for failure in this area we may often refrain from

---

[1] For one influential example, see Tooley 1972, 40.
[2] See Sieghart 1985, ch. 8.
[3] Though we would say that governments are obligated not to distribute these rights in a discriminatory fashion.

labelling such failure a dereliction of duty or an injustice. Thus rights in this second category may lack the conceptual connections with the notions of obligation and justice, and thus also the distinctive cutting edge, of rights in the first category. Speaking of economic, social, and cultural rights will often have the effect not of defining requirements of justice for individuals considered separately but of establishing ideals or objectives for societies considered collectively.

The coexistence of these two categories of rights in international law stands as a fossil record of the expansion, and consequent dilution, of the notion of a right. Rights in their narrow sense formulate urgent or insistent demands precisely because they constrain the pursuit of social goals. They are thus completely at home when they are invoked to protect basic liberties, due process, or political participation, since we are prepared to say that a society must secure these goods for all of its citizens alike regardless of any further goals it might elect to pursue. However, a society which fully satisfies this standard might for all that still contain widespread poverty, illiteracy, unemployment, or disease. When we notice this fact the obvious remedy is to formulate additional standards to eliminate these further economic, social, and cultural evils. And when we wish to underscore the importance of these additional standards the obvious device is to formulate them in the same language of rights. Thus by degrees do we move from using rights to impose constraints on the pursuit of social goals to using them to formulate just such goals.

If we wish to arrest this conceptual drift then we must construct a concept of a right which preserves something of its original normative function. Thus far, however, we have only a vague characterization of that function in terms of constraining the pursuit of goals. In order to sharpen that characterization we need to develop richer accounts both of a constraint and of a goal. We will postpone the latter task until we come to consider moral theories which treat goals as basic. The aim of this chapter, therefore, is to explicate the special kind of constraint imposed by rights.

Although our destination is an adequate concept of a moral right, we will do well to focus initially on a more tractable case. It is a commonplace that rights admit of different varieties—legal, customary, moral, etc. I propose to adopt the working hypothesis that these varieties are related as the several species of a common genus. Although this assumption is questionable (and will later be questioned) it seems the most natural one to make at the outset of our inquiry. If the hypothesis is correct then all species of rights must share some common concept of a right, their differentiae being given by their modifiers—'legal', 'customary', 'moral', etc. But then we ought to be able to uncover this common concept by examining any of the particular species. Because moral rights appear to have some especially puzzling features it seems reasonable to select some other category as our initial specimen. It is also a commonplace that some rights are bestowed on us by conventional rule systems which apply to us. If we can come to understand the nature of these rights then we may be able to use this understanding to advantage in addressing the more problematic case of moral rights. A two-stage conceptual strategy therefore appears to be promising: first we isolate the concept of a right implicit in the case of conventional rights and then we extend or extrapolate it to the case of moral rights. Our first task, therefore, is to explore the role of rights in conventional rule systems.

2.1 BUILDING BLOCKS

Where the analysis of rights is concerned, the beginning of wisdom, though not the end, lies in Wesley Hohfeld's celebrated classification of 'fundamental legal conceptions'.[4] Hohfeld complained that the notion of a right as it figured in judicial reasoning was multiply ambiguous and that these ambiguities encouraged legal arguments to equivocate as they slid effortlessly from one sense of the notion to another. His remedy for this conceptual muddle was to map the logical relations among a set of 'fundamental conceptions' so as not only to distinguish rights from the other items with which they were commonly confused but also to show how more complex legal notions, such as that of a trust, could be constructed out of these simple elements.

Despite its many virtues, however, Hohfeld's analytic work suffers from a number of limitations of its own. While many of these are matters of relatively insignificant detail which can be remedied during the course of our discussion, some require mention

---

[4] Hohfeld 1919.

at the outset. To a philosopher's eye the most obvious of them is Hohfeld's failure to analyse any of his conceptions. Instead, he contented himself with organizing them into two tables of opposites (contradictories) and correlatives (equivalents). Each notion was therefore given a contextual definition by virtue of being connected with its cognates, as well as an ostensive definition by virtue of abundant illustration from judicial argument, but Hohfeld betrayed no curiosity about whether his conceptions could themselves be reduced to some set of more primitive notions. Indeed, by calling them 'fundamental', '*sui generis*', and 'the lowest common denominators of the law' he implied that they were themselves primitive.[5] If this is what he believed then he was mistaken. As we shall see, it is possible to construct all of Hohfeld's conceptions out of two kinds of modality plus a few accessory items. Nor is the undertaking of this reductive exercise a mere matter of logical fussiness, since it will enable us both to display the logical structure which underlies Hohfeld's tables of opposites and correlatives and to reveal some patterns among the items in those tables which escaped his attention.[6]

If the first limitation of Hohfeld's analysis lies in its lack of a systematic basic vocabulary, the second lies in one of its principal conclusions. Hohfeld contended that a right 'in the strictest sense' or in its 'limited and proper meaning' was itself one of his 'fundamental conceptions', namely the correlative of a duty owed to some second party.[7] He was therefore committed to holding (1) that the notion of a legal right has but one strict or proper sense and (2) that this sense is (in his scheme) atomic rather than molecular in its structure. Each of these contentions is, however, dubitable. If (1) is interpreted as a report about the actual state of judicial reasoning then it is disconfirmed by Hohfeld's own complaint about the multiple ambiguity of rights in such reasoning and by the mass of evidence he assembled in order to support that complaint. If on the other hand it is interpreted as a stipulation, as it was doubtless intended, then Hohfeld provided virtually no argument in favour of it and against the contrary view that importantly different notions

---

[5] See, for example, ibid. 36, 64.
[6] Other analyses of Hohfeld's categories may be found in Anderson 1962; Kanger and Kanger 1966; Fitch 1967; Ross 1968, ch. 5; Pörn 1970; Lindahl 1977; Robinson *et al.* 1983; Stoljar 1984, ch. 5; Wellman 1985, ch. 2.
[7] Hohfeld 1919, 36, 38.

of a right, each of them suitably strict and proper, might profitably coexist in the law. Nor, if we accept (1), did he offer any reason for thinking that the sole strict and proper notion was one of his 'fundamental conceptions' rather than a bundle of related conceptions. Hohfeld thus arbitrarily closed some important questions about the univocity and complexity of rights which it will be important to our enterprise to leave open, at least at the initial stage of deploying a basic vocabulary. In order to leave these questions open we must alter Hohfeld's terminology in one important respect. In common with most of the subsequent literature, and in line with Hohfeld's own suggestion, I shall call the correlative of a duty owed to some second party not a right but a claim.[8]

For our purposes Hohfeld's treatment has a third shortcoming which, however, it would be unfair to represent as a fault in it. Hohfeld was concerned solely with legal conceptions or, as he also called them, jural relations, and thus he was concerned solely with legal rights. By contrast, at this stage of our inquiry we are seeking an account of the internal structure not just of legal rights but of all forms of conventional rights. In order to construct this account we will have to expand Hohfeld's narrow focus on rights within a municipal legal system to include rights within any conventional rule system, legal or non-legal. Hohfeld's distinctions among types of legal conceptions presuppose logically prior distinctions among types of legal rules and are unintelligible except against this background. In general, it seems that any adequate conceptual analysis of conventional rights is possible only if we attend to the different forms which conventional rules can take and the different functions which they can have. Hohfeld's analysis can be understood as an attempt to classify just these forms and functions. But his attempt was limited to one special case of a conventional rule system, namely a municipal legal system. Our range must be wider.

Because of these three limitations Hohfeld's analytic framework will not serve our purposes as its stands. Thus what follows should be construed not as an explication of that framework but rather as a reconstruction of it which departs from Hohfeld on many issues of terminology, methodology, and substance. For all of these departures, however, it retains and supports a great many of Hohfeld's contentions, including his basic commitment to two

[8] Ibid. 38, 71.

independent sets of conceptions, the members of each of which are
connected by logical relations of contradiction and equivalence.
More than that, however, it remains Hohfeldian in spirit and
inspiration, since it shares Hohfeld's conviction that the nature of
rights can be illuminated only by developing and deploying a
clearly articulated normative vocabulary.

We begin by imagining the existence of a system of rules applying
to a set of agents. We may imagine these rules as the municipal legal
system of some state, or as the rule system of some institution (a
corporation, a church, a university, a labour union, a club, etc.), or
as the rule system of some less formal association (a family, a team,
an interest group, a gang, a commune, etc.), or as any other set of
rules which satisfies the very loose conditions of forming a system
and governing the conduct of a set of agents. Furthermore, we need
not insist that the system contain only what we would ordinarily
call rules as opposed to laws, by-laws, statutes, decrees, directives,
edicts, ordinances, standing orders, regulations, injunctions, mandates,
norms, precepts, guidelines, canons, principles, or whatever. We
will use the term 'rule' indifferently to embrace all of these items,
the distinctions among them being unimportant at this stage of our
inquiry. We will also not pause to ask what a rule is, or what it is
for a set of rules to form a system, or what it is for a system of rules
to be conventional, or what it is for a given system to apply to a
given set of agents. All of these are complex and difficult questions
and we will need to answer at least some of them in the next
chapter when we turn to constructing existence conditions for
conventional rule systems, and thus for the rights which they
confer. For the moment, however, it will suffice to rely on our
intuitive familiarity with actual instances of rule systems which
apply to (more or less determinate) sets of agents.

We next imagine that among the various sorts of rules which this
system contains some have the function of regulating or directing
the conduct of the agents to whom they apply. One way in which
they accomplish this function is to stipulate what those agents must
do or are required to do or, alternatively, what they must not do or
are forbidden to do. Familiar examples of rules imposing requirements
include statutes obliging us to file income tax returns, traffic
regulations telling us to turn our headlights on at dusk, social rules
dictating that we stand during the playing of the national anthem,
and the like. But we are perhaps more accustomed to rules which

direct conduct not by telling us what we must do but by telling us what we must not do—not, that is, by prescribing but by prohibiting. The provisions of the criminal law ordinarily take this form, as do many of the regulations found in other codes of conduct—'Employees must not use company cars for private purposes', 'Smoking prohibited in this compartment', 'No parking between midnight and 7.00 a.m.', etc. However, although there are important pragmatic differences between rules which prescribe conduct and rules which prohibit it, the logic of both is identical. From a purely semantic point of view a rule of the form 'Blanks are required to bleep' can always be trivially reformulated as 'Blanks are forbidden not to bleep', while 'Blanks are forbidden to bleep' can likewise be rewritten as 'Blanks are required not to bleep'. Prescribing an act is logically equivalent to prohibiting its omission, while prohibiting an act is equivalent to prescribing its omission. In order to formulate both prescriptions and prohibitions we need (besides expressions for agents and acts) only a single deontic notion (either 'required' or 'forbidden') plus negation.

What prescriptions and prohibitions have in common is that they impose restrictions, or constraints, on those to whom they apply. The constraints are not, of course, literal. Rules do not constrain in the way that ropes or bars constrain, and it is usually physically possible to break them. But the analogy between a physical and a normative constraint is instructive. Just as the former reduces the range of options physically possible for me (i.e. compatible with the laws of nature), the latter reduces the range of options deontically possible for me (i.e. compatible with the rules of conduct). The analogy between the two kinds of constraints serves to remind us that deontic categories (required/forbidden) are counterparts, or perhaps special cases, of alethic modal categories (necessary/ impossible).

It also serves to remind us of a third category, namely an act which is permitted (deontically possible). Since an act is permitted in this sense just in case it is not forbidden, and therefore just in case its omission is not required, this third deontic category can readily be defined in terms of either of the other two (plus negation). But we can equally define both of them in terms of it: an act is forbidden just in case it is not permitted and required just in case its omission is not permitted. From a logical point of view the members of the deontic triad (required/forbidden/permitted) are all interdefinable and none is more primitive than the others. This symmetry reflects the like symmetry of the alethic modalities (necessary/impossible/possible).

Since the prohibitions contained in most rule systems are relatively specific, and since an act is permitted if it is not prohibited, most rule systems contain a great many implicit permissions. Further, since an act is permitted if it is required the most common way for rule systems to permit acts explicitly is by prescribing them. But again these features of rule systems are products of pragmatics rather than semantics. It is quite possible to imagine a rule system consisting of a perfectly general prohibition followed by a list of specific exceptions.[9] In such a system every specific rule would be permissive rather than restrictive and any act not explicitly permitted would be forbidden. Actual rule systems tend to be mixtures of restrictive and permissive rules, where the latter ('Smoking permitted in this area', 'Visiting hours from noon to 8.00 p.m.', 'Use of calculators allowed in this examination') generally formulate exceptions to (implicit or explicit) background prohibitions.

The simple deontic vocabulary which enables us to formulate restrictive and permissive rules also enables us to begin assembling the first set of Hohfeld's 'fundamental conceptions'. For reasons to be explained shortly, we will call these first-order conceptions. Although from a semantic point of view we could start either with restrictions (whether prescriptions or prohibitions) or with permissions, there are pragmatic reasons for favouring the former. Since in the absence of all rule systems everything would be (implicitly) permitted, the primary function of such systems is to impose constraints on this condition of unlimited permissiveness. It is therefore natural to treat restrictions as pragmatically prior to permissions. Where a rule is restrictive we may say that it imposes a duty on those to whom it applies: a positive duty if the rule has the form of a prescription, a negative duty if it has the form of a prohibition.[10] Thus I have a duty to attend the meeting just in case some rule requires me to do so, and I have a duty not to disrupt the meeting just in case some rule forbids me to do so. It should be

---

[9] See Raz 1975, 85–97. My distinction between restrictive and permissive rules coincides with Raz's distinction between mandatory and permissive rules.

[10] In the literature restrictive rules are often called duty-imposing rules, as in Hart 1961, chs. 3 and 5.

noted that this simple deontic notion of a duty is considerably broader than our ordinary notion. In actual rule systems duties are often distinguishable from such cognate items as obligations, responsibilities, expectations, tasks, and the like, and it can be quite counter-intuitive to treat all of the requirements imposed by such systems as duties in the strict sense.[11] But just as we are using the term 'rule' in an artificially broad sense which abstracts from the differences among particular rule systems, we will use the term 'duty' in a correspondingly broad sense. Since we want to reveal the common structure of rights in all rule systems we need some notion which will cover any case of being obliged by a rule either to do or to omit some act, and the notion of a duty seems best suited to this role.[12]

Although Hohfeld treats duties as one of his 'fundamental conceptions', his notion of a duty is different in a crucial respect from the simple deontic notion which we have thus far defined. All of Hohfeld's conceptions are relations between two distinct parties. Thus each of Hohfeld's duties has not only a content (what it is a duty to do) and a subject (the party whose duty it is) but also an object (the party to whom the duty is owed). His duties are, we might say, directional. The simple deontic notion of a duty, by contrast, is non-directional and thus non-relational. If we wish to generalize Hohfeld's conception of a legal duty to any conventional rule system we must add some further ingredient which captures the directionality of relational duties. Unfortunately it is impossible to decide at this stage of our inquiry just what this further ingredient should be. The literature on relational duties offers two dominant accounts of what it means for a duty to be owed to some specified person or set of persons. For strategic reasons it will be best to postpone consideration of these accounts until we have developed more of our Hohfeldian vocabulary. In the meantime, therefore, we will simply place on the record our need for the futher ingredient and leave its nature unspecified.

Relational duties are crucial to our enterprise because rights cannot be understood without them. In developing the notion of a relational duty we are not, however, committing ourselves to Hohfeld's assumption that all duties, or even all legal duties, are

---

[11] For some of the ordinary-language nuances of 'duty' and 'obligation' see White 1984, chs. 3 and 4.

[12] Henceforth I shall freely interchange 'duty' and 'obligation'.

relational. Whether we share that assumption will depend on how we come to interpret the directionality which is essential to such duties. Whether we share it or not, however, relational duties will remain central to our analysis of the nature of rights. Furthermore, since it is not a matter of dispute that some duties are relational we can safely focus on this category without prejudice to the issue of whether it is the only category. We also need not share Hohfeld's assumption that the subject and object of a relational duty must be distinct individuals. At this stage of our inquiry we must leave open the possibility that some duties might be held by collectivities, or owed to collectivities, or both held by and owed to the same party.[13]

Once we have this notion in hand then it is a simple matter to define its Hohfeldian correlative and opposite. Hohfeld's correlatives are normative positions linked by one of his normative relations; if his conceptions were not relations then they would admit of no (logical) correlations. The correlative of a relational duty is a claim, which is simply the duty described from the vantage point of its object. Thus I have a claim against you that you feed my cat just in case you have a duty to me to feed my cat; the directionality of the claim hooks up with that of the duty. There is a straightforward semantic formula for converting a relational duty into a claim, and vice versa. Whichever item you begin with you can generate the other just by reversing the subject and object and leaving the content unchanged. Because the content of a duty can only be an act of the duty-bearer, it follows that the content of a claim cannot be an act of the claim-holder. I cannot have claims *to do*, only claims *that others do*.[14]

Hohfeldian correlatives are logical equivalents with different subjects. Hohfeldian opposites, by contrast, are logical contradictories with the same subject. The Hohfeldian opposite of a relational duty is a liberty.[15] In order to identify the opposite of a duty we must negate its content, thus converting a duty to do into a duty not to

---

[13] For expository simplicity I shall assume in what follows that the parties to a relational duty are distinct individuals. This assumption will be questioned in ch. 7.

[14] Kanger and Kanger 1966 depart from tradition by treating the contents of Hohfeldian relations not as actions but as conditions or states of affairs; they are followed in this by Lindahl 1977. However, the difference between the two conventions is merely nominal, since Kanger and Kanger are compelled to introduce the notion of causing a state to be the case, or seeing that it is the case.

[15] I follow common practice here in substituting 'liberty' for Hohfeld's 'privilege'.

do (or vice versa). The liberty which is the opposite of a duty then has the same subject and object as the duty but takes this negated content. Thus my liberty with respect to you to play my stereo is the opposite of my duty to you not to play it. Because the subject of a liberty and its opposite duty are the same, and because the content of a duty can only be an act of the duty-bearer, it follows that the content of a liberty can only be an act of the liberty-holder. I cannot have liberties *that others do*, only liberties *to do*.

Since Hohfeldian liberties are the opposites of relational duties they are themselves relational. A liberty must have both a subject (the subject of the opposite duty) and an object (the object of the opposite duty). It is therefore possible for me to have a liberty with respect to one person but to lack a liberty of the same content with respect to another. If, for instance, I have no duty to you not to play my stereo but I do have such a duty to my room-mate then I have a liberty with respect to you to play it but no such liberty with respect to her. Out of this relational notion of liberty we can, of course, easily construct a non-relational notion: I have a liberty *tout court* to do something just in case I have a liberty with respect to everyone to do it, thus just in case I have no duty to anyone not to do it.

Just as the simple deontic notion of a duty differs from our ordinary notion, so this notion of a liberty differs in some respects from our ordinary one. It should not, of course, be confused with physical liberty, though freedom from a deontic constraint is the normative counterpart of freedom from a physical constraint. The main departure from ordinary usage lies in the fact that the Hohfeldian liberty does not guarantee any choice among alternative options, since having the liberty to do something is compatible with lacking the liberty not to do it. This feature of liberties follows from the fact that they are simply deontic permissions: I am at liberty to attend the meeting just in case the rules permit me to do so. If we assume that a rule system will not both require and prohibit the same act then if I am required to attend the meeting I am also permitted to do so. It follows that I have a liberty to do whatever I have a duty to do, though I do not have a liberty not to do it.

Since the Hohfeldian notion of a liberty is well suited to our conceptual purposes, this counter-intuitive feature is not sufficient inducement to alter it. In any case we can easily generate the more familiar notion by utilizing a terminological suggestion made by

Joel Feinberg.[16] Suppose that I have no duty either to attend the meeting or not to do so. I thus have two logically distinct Hohfeldian liberties. Call each of these a half liberty and their conjunction a full liberty. Then I have a half liberty to attend the meeting, a half liberty not to attend it, and a full liberty to attend it or not. In general I have a full liberty with respect to anything which I am neither obligated to do nor obligated not to do. Unlike half liberties, full liberties ensure a normatively unencumbered choice between options. To the extent that we prize the opportunity for choice we will obviously prize full liberties more than (mere) half liberties, and we should therefore not be surprised if full liberties play an important role in our analysis of rights. None the less, for the present all liberties not explicitly identified as full may be assumed to be half liberties.

We are now in a position to display the logical connections among first-order Hohfeldian normative relations. Where X and Y are persons and V is some act, the rows in the following matrix give correlatives and the columns (and diagonals) give opposites.

X has a liberty with respect to Y to V	Y has no claim against X that X not V
X has a duty to Y not to V	Y has a claim against X that X not V

Because of the connections among these relations, a change in any one of them is automatically a change in all the others. However, the rule system which we have thus far imagined is entirely static: it contains rules which determine the first-order relations of those to whom the system applies, either by imposing duties or by conferring claims and liberties, but it contains no mechanism for creating, altering, extinguishing, or otherwise manipulating these relations. For this purpose we will need a further set of rules with a different form and function.

The rules which enable us to manipulate normative relations must be of a higher logical order than the rules which define those relations. Thus it is now obvious why we have called our initial set of Hohfeldian conceptions first-order relations. Our next set of conceptions will consist of second-order operations on those

---

[16] Feinberg 1980, 157, 237. The same distinction is drawn in terms of unilateral and bilateral liberties in Hart 1982, 166–7.

relations.[17] The rules which define these operations have the function not of regulating or directing conduct but of enabling or facilitating it.[18] They therefore determine not what we may do but what we can do, not what is permissible but what is possible. Just as the deontic modalities (required/forbidden/permitted) provided the materials for constructing our first-order relations, the alethic modalities (necessary/impossible/possible) will provide the materials for constructing our second-order relations.[19]

Since none of the alethic modalities is logically prior to the others then once again we could in principle select any of them as basic. However, the same considerations which earlier led us to assign pragmatic priority to rules which impose constraints will now suggest assigning the same priority to rules which confer abilities. In the absence of all rules assigning first-order relations everything is permitted, and thus the novel contribution of such rules is to impose restrictions. By contrast, in the absence of all rules enabling us to manipulate such relations no changes in them are possible, and thus the novel contribution of such rules is to facilitate such changes. Thus where first-order rules are concerned the required and the forbidden are pragmatically prior to the permitted, while where second-order rules are concerned the possible is pragmatically prior to the necessary and the impossible.

Our first step will therefore be to define a Hohfeldian power. Basically, I have the power to affect (that is, to alter or sustain) some normative relation just in case the rules of the system make it possible for me to do so. A rule which confers a power thus creates the normative analogue of a physical ability. Familiar instances of

---

[17] These conceptions can, and should, be defined more broadly as operations on any normative positions, whether relational or non-relational, first-order or higher-order. For simplicity, however, I will confine myself to operations on first-order relations.

[18] The two types of rules are called by Hart and Raz duty-imposing and power-conferring; see Hart 1961, chs. 3 and 5; 1982, ch. 8; Raz 1975, chs. 2 and 3. I have avoided this terminology because not all rules which regulate conduct impose duties (some confer liberties), while not all rules which facilitate conduct confer powers (some impose disabilities). Ross draws the same distinction in terms of norms of conduct and norms of competence; see Ross 1968, ch. 5.

[19] The distinction between deontic and alethic modalities is used to draw the distinction between first- and second-order Hohfeldian relations in Anderson 1962; Fitch 1967; Pörn 1970, 52 ff.; and Lindahl 1977; 206–10. The latter relations are claimed to be reducible to the former in Kanger and Kanger 1966; Ross 1968, 118–20; and Lindahl 1977, 212 ff. Nothing in the functional distinction between the two orders turns on this reducibility issue.

powers in rule systems include the power of governing bodies to change the system by creating new rules or abolishing old ones, the authority of adjudicative bodies to alter the normative relations of individuals by applying the rules of the system to them, the capacity of private individuals to alter their own normative relations by making agreements, and so on. It is the existence of rules conferring powers of these sorts which renders a rule system dynamic rather than static, by investing those subject to the system with some degree of control over its constraints.

Pragmatically speaking, powers are the second-order counterparts of duties. Just as the novel contribution of first-order rules consists of restrictions, the novel contribution of second-order rules consists of abilities to manipulate those restrictions. But logically speaking, powers are the second-order counterparts of liberties. Just as each of my liberties specifies something that I may do (that is permissible for me), so each of my powers specifies something that I can do (that is possible for me). It is thus possible for me to do what I lack the liberty to do; any such act will be a violation of the rules. But it is not possible for me to do what I lack the power to do; any such attempt will be, as the lawyers say, a nullity. Furthermore, just as the content of a liberty must be an act of the liberty-holder so the content of a power must be an act of the power-holder, and just as liberties can be either half or full so can powers. I have a half liberty to do something when the rules permit me to do it, and this is compatible with having no liberty not to do it. Similarly, I have a half power to alter some normative relation when the rules enable me to do so, and this is compatible with having no power not to alter it. I have a full liberty only when I may either do or not do something, and I have a full power only when I can either alter or not alter some relation.

Again like liberties, powers are not usually thought of as relations. The best way to bring out the relational aspect of a power is to define the further Hohfeldian conception of an immunity. Basically, I have an immunity against having some normative relation affected just in case the rules of the system make it impossible to affect that relation. Possession of an immunity by one person thus ensures the lack of a power on the part of others. But it is quite possible for me to have an immunity against your affecting some relation of mine without having the same immunity against someone else. Suppose, for instance, that I have offered to purchase

my room-mate's piano for two hundred dollars. In virtue of my
offer she now has the power to impose a duty on me to pay her two
hundred dollars, which power she can exercise by accepting the
offer. I have an immunity against your imposing on me the duty to
pay her two hundred dollars, but no such immunity against her
doing so. Thus an immunity, like a claim, has an object, namely the
person against whom it is held. Of course it is possible for me to have
an immunity *tout court*, if some normative relation of mine is
immune to alteration by anyone else.

The example which shows that immunities are relational also
shows that powers are relational. While my room-mate has the
power to impose on me a duty to pay her two hundred dollars she
has no such power over you. Thus a power also has an object,
namely the person over whom it is held. Rule systems commonly
confer powers with (more or less determinate) objects, as in the case
of the authority of parents over their children, or of the executive
committee of a club over its membership. But once again it is
possible to have a power *tout court*, as when your acquiring a piece
of property imposes upon everyone alike the duty not to trespass on
it.

With the notions of a power and an immunity in hand, we can
map the logical connections among the various second-order
Hohfeldian relations. Where X and Y are persons and R is some
normative relation, once again the rows in the following matrix
give correlatives and the columns (and diagonals) give opposites.

X has a power over Y to affect R	Y has a liability to X's affecting R
X has a disability with respect to Y to affect R	Y has an immunity against X's affecting R

If we compare the matrices for the two sets of relations then it
becomes possible to uncover some symmetries between them. We
have already indicated the respects in which powers are the second-
order counterparts of liberties. But now it becomes apparent that
disabilities must be the second-order counterparts of duties. This
symmetry is obscured by the pragmatics of the two levels. Where
first-order rules are concerned it is natural to treat duty as the
positive notion and liberty as its privation. But in order to do this
we need to specify that the opposite of a duty to do something is a

liberty not to do it. By contrast, where second-order rules are concerned it is natural to treat power as the positive notion and disability as its privation, in which case no negation of content is necessary. The underlying symmetry between duties and disabilities emerges nicely if we retreat to a more primitive vocabulary. I have the liberty to do something just in case doing it is permissible for me, and I lack the liberty just in case doing it is impermissible for me. Likewise, I have the power to affect some relation just in case affecting it is possible for me, and I lack the power just in case affecting it is impossible for me.

It should now be clear that immunities are the second-order counterparts of claims. Like a claim, an immunity can take as its content only an act ascribed to its object rather than its subject. Furthermore, just as a claim held by one person necessarily limits the liberties of the person who is its object, an immunity held by one person necessarily limits the powers of the person who is its object. Claims and immunities thus both impose constraints (of different sorts) on others. Hohfeld thought that a power 'bears the same general contrast' to an immunity that a claim does to a liberty, and that powers and claims are analogues.[20] But in this he was attending not to the semantics of his relations but to their pragmatics. In their logic a power stands to an immunity as a liberty stands to a claim, and it is powers and liberties which are analogues.[21] Thus are patterns obscured by the lack of an adequate vocabulary.

The distinction between rules defining first-order normative relations and rules defining second-order operations on those relations can easily be iterated to higher levels. Thus a rule system may contain rules which facilitate operations on second-order operations (powers over powers), rules which regulate the exercise of second- and higher-order operations (duties or liberties to use powers), rules which facilitate operations on these still higher-order normative relations, and so on. However, further iterations reveal no conceptual novelties, and the two orders of relations which we have defined will do nicely as the basic materials out of which rights can be constructed.

[20] Hohfeld 1919, 60.
[21] See Fitch 1967, which gets the analogies right.

2.2 CONCEPTUAL ALTERNATIVES

Thus far we have followed Hohfeld in mapping the logical relations among eight normative positions. Our next task is to use these materials to assemble a concept of a right. We may begin by reducing our inventory a little. Consider our four pairs of correlatives. Each pair consists of two normative positions which are logically connected by virtue of being equivalent descriptions of the same underlying relation. Within each pair one of these positions is a normative advantage, the other a normative disadvantage. Thus, for example, in the claim–duty pairing holding a claim is standardly a benefit, bearing a duty standardly a burden. Similarly, having a liberty is standardly more advantageous than lacking a claim, and having a power or an immunity likewise more advantageous than having a liability or a disability. In singling out liberties, claims, powers, and immunities as normative advantages we are not of course denying that possession of them can make us worse off on particular occasions, nor that possession of a normative disadvantage can make us better off. Instead, we are pointing to the fact that occupying one of these positions means having the rules of the system in question on one's side *vis-à-vis* the occupant of the correlative position. It is thus to be so situated by the rules as to be able to prevail over the other in some specified conflict. Since normative opportunities can be squandered, or deliberately foregone, they will not always deliver resultant benefits. But they will always put one in a position to benefit.

Whatever rights may be, everyone agrees that they too are normatively advantageous. Thus we may limit the materials out of which they are to be constructed to those Hohfeldian positions which are advantages. But which of them? Here we can broadly distinguish two hypotheses. One is that rights are simple, thus that every right consists of just one position. The other is that rights are complex, thus that every right consists of some bundle of different positions. These two hypotheses do not, of course, exhaust the possibilities; some rights might be simple while others are complex. Further, we can easily distinguish between monistic and pluralistic versions of each hypothesis. A monistic version of the first hypothesis would hold that every right consists of the same normative position, while a pluralistic version would allow different rights to consist of different positions (though only one in each case). Likewise, a monistic version of the second hypothesis would hold that every right consists of the same bundle of positions, while a pluralistic version would allow different rights to consist of different bundles. Given the variety of possible analyses, it will be out of the question to give a full hearing to each. Thus I will confine myself to offering some considerations in favour of a view which is both complex and pluralistic.

Let us begin, however, with the idea that rights are simple, thus that every right is identical to some single Hohfeldian advantage. Which one (or ones)? As we have seen, Hohfeld himself identified rights 'in the strictest sense' with claims, thus holding a monistic version of the simple view. Certainly it seems likely that if all rights are to be identified with the same simple position then claims will have, as it were, the strongest claim. But it will be instructive to postpone consideration of Hohfeld's proposal a little, until we have had an opportunity to examine the credentials of the other contenders. Could rights, then, be just liberties? Hobbes thought so:

For though they that speak of this subject, use to confound *Jus*, and *Lex*, *Right* and *Law*; yet they ought to be distinguished; because RIGHT, consisteth in liberty to do, or to forbeare; whereas LAW, determineth, and bindeth to one of them: so that Law, and Right, differ as much, as Obligation, and Liberty; which in one and the same matter are inconsistent.[22]

It is immediately noticeable, however, that Hobbes identified the right to do something not with the mere (half) liberty to do it but with the (full) liberty 'to do, or to forbear'. Technically, therefore, he held that every right is a bundle of two related Hohfeldian positions, thus defending a (monistic) version of the complex view. Still, since both of the positions in question are liberties, it is probably more illuminating to treat Hobbes's analysis of rights as a (slightly deviant) simple one.

Certainly if rights are just liberties then there is little plausibility in the idea that they are mere half liberties. One element present in at least many rights is the agent's opportunity to choose, free of deontic constraints, among alternative courses of action. Since a half liberty to do something is compatible with, because entailed by, a duty to do it, it cannot by itself capture this feature of deontically unhindered choice. Thus at least some rights either are, or at least contain, full liberties. Indeed it is questionable whether

[22] Hobbes 1968, 189.

any right can consist of only one liberty without its complementary half. Imagine that some rule system imposes on you the duty to bleep. Knowing this, would I then want to say that you have a right to bleep? Ordinarily, conventions of conversational implicature would dictate otherwise. In saying that you have a right to bleep I conversationally imply that I have given a full account of your normative position with respect to bleeping, thus that you do not also have a duty to bleep. Saying that you have a right when I know that you have a duty would then be akin to saying that you have two children when I know that you have three. However, if it would be conversationally deviant in this latter case for me to say that you have two children this fact does nothing to show that what I say is false; after all, if you have three children then you do also have two. Likewise, even if it would be conversationally deviant for me to say that you have the right to bleep, you may have the right for all that. And in some cases it is not even conversationally deviant. Thus, for example, I might quite intelligibly say that you have the right to chair the meeting when I know that you have the duty to do so. In such instances of what Feinberg has called 'mandatory rights' the background presupposition is generally either that the duty in question is a mark of some desirable rank or privilege (as in the case of the right and duty to vote or to serve as a juror, both of which can be regarded as privileges of citizenship) or that it can otherwise be seen as a benefit as well as a burden (as in the case of a child's right and duty to attend school).[23] For all we have said thus far, there might be good reasons for thinking that the notion of a mandatory right is deviant or oxymoronic. But at this stage of our inquiry it would be arbitrary to insist that rights must contain full liberties.

The special case of mandatory rights is, however, instructive in one respect. Where we are tempted to ascribe such rights there are normative elements in the picture besides the right-holder's duty and consequent half liberty. If you have a genuine right to chair the meeting then the other members are obliged not to interfere with your doing so. Likewise, if I have a legal right to vote then private individuals are required not to prevent me from doing so (e.g., by impeding my access to the polling station) and electoral officials are

<hr>

[23] See Lyons 1970, 55; Feinberg 1980, 157–8, 232 ff.; Hart 1982, 173–4. Mandatory rights are claimed to be illusory in Robinson *et al.* 1983, 289–90.

required to assist me in doing so (e.g., by offering me a ballot). A similar framework of duties imposed on others is visible in each of the other cases of mandatory rights. This fact suggests something important, namely that a mere half liberty by itself, without any supporting duties of assistance or non-interference imposed on others, does not constitute a right. And if this is so, then the same seems to hold of full liberties. Hobbes's natural rights, which are fully enjoyed by everyone in the state of nature, are notoriously lacking in such supporting duties. My possession of them therefore imposes no normative constraint on others. But if the choice which is provided by a full liberty seems to be one ingredient in our ordinary concept of a right, another such ingredient seems to be the imposition of some constraints on others. As we have seen, this ingredient, however vague it might be, is especially prominent in the case of those rights which play a role in moral/political argument. If rights are to constrain others, however, they must contain something more than mere liberties (whether half or full). The view that rights are full liberties nicely captures the idea of deontically unencumbered choice which is central to at least some rights. But it omits the idea of insulation against the liberties of others which is just as central. In order to capture this latter idea liberties must be surrounded by what H. L. A. Hart has usefully called a 'protective perimeter' of duties imposed on others.[24] But the structure of a right which we assemble in this manner will be necessarily complex.

Since powers are the higher-order counterparts of liberties the proposal that (at least some) rights are just powers requires little independent assessment. As in the case of liberties, we can distinguish half powers and full powers. The counterpart of a mandatory right will therefore be something like a 'necessary power'—i.e., the power to affect some normative relation in some particular way without the power not to do so, or vice versa.[25] Just as we can question whether mandatory rights are genuine rights, we could also question whether necessary powers are genuine rights. But anyone who chose to identify rights with powers in the first

---

[24] Hart 1982, 171–3.
[25] I have a necessary power whenever some particular alteration of a relation is impossible for me. I then lack the power to alter it, but (trivially) have the power not to alter it.

place would be thinking of those cases in which someone has both the power to affect some relation and the power not to do so. The holder of such a (full) power has the alethic analogue of deontic freedom of choice. If I have a full liberty then there are different options among which I may choose (which are permissible for me). If I have a full power then there are different options among which I can choose (which are possible for me). Thus both full liberties and full powers are capable of being *exercised*, since in each case exercising them means selecting one of the available (permissible/possible) options.

To the extent that (deontically or alethically) unencumbered choice is one important ingredient in our common conception of a right, then the idea that at least some rights are simply full powers will hold some attraction. Thus we might think that my right to make a will consists solely of my power (or capacity) to do so or not as I choose, or that a police officer's right to make an arrest consists solely of her power (or authority) to do so or not as she chooses. But a little further reflection reveals that in such cases, just as in the parallel cases of full liberties, we invariably attribute to the right-holder some additional normative advantages. The most obvious of these is the one which we added to liberty, namely a protective perimeter of duties borne by others. Thus we would not think much of my 'right' to make a will if, while I have the power to do so or not as I choose, others are permitted to prevent me from exercising this power. A mere power, which by itself can impose no constraints on others, seems too defenceless to count as a genuine right.

In the case of powers, however, there are two further ingredients whose addition to the package is necessitated by the fact that powers are second-order normative positions. The first is an immunity against the like powers of others. We would also not think much of my 'right' to make a will if, while I have the power to do so or not as I choose, others have the same power (to make a will for me or not as they choose). We ordinarily assume that my right to make a will confers on me exclusive control over whether anything will come to count as *my* will. But my (full) power to make a will does not by itself assure me of this control. To do so it must be accompanied by a guarantee that no one else has a power with the same content. Thus the protective perimeter of powers consists not just of claims (i.e., duties borne by others) but also of immunities (i.e., disabilities borne by others). Since immunities are the higher-order counterparts of claims, this is exactly what we should expect. Meanwhile, the second additional ingredient is one which we need not add to a liberty because it is itself a liberty. We would also not think much of my 'right' to make a will if, while I have the power to do so or not as I choose, I also have the duty not to do so. Thus we must add to a power at least the higher-order liberty to exercise it. Must we also add the liberty not to exercise it? If I am to have full control over whether anything will come to count as my will, then my full power must be accompanied by a full liberty to exercise it or not as I choose. But just as we can make some sense of first-order mandatory rights (i.e., mandatory or directed liberties), so we can also make some sense of second-order mandatory rights (i.e., mandatory or directed powers).[26] Thus, for instance, a university registrar may be empowered to issue academic transcripts but also required to do so in the case of any applicant who satisfies some specified conditions. It would not be a gross abuse of language to say that the registrar has the right to issue transcripts, despite the fact that (when these conditions are satisfied) he lacks the liberty not to.

Whatever we may think of second-order mandatory rights, it is clear that when we think of powers as rights we are normally thinking of them as ingredients in some complex package. Interestingly, the same need not hold for immunities. We are less tempted to identify rights with immunities if only because immunities tend to be both less common and less visible than other normative positions. But some high-profile rights are arguably just immunities. Consider, for example, a constitutionally guaranteed right of free speech.[27] The normative effect of entrenching such a right in the constitution of a municipal legal system might be to disable the legislature from regulating speech in certain ways (e.g., on grounds of content rather than occasion). If so, then the right confers an immunity against legislative restriction of a first-order liberty (or bundle of liberties). The right therefore entails the liberty, which might seem to show that once again it is a package consisting at least of an immunity plus a liberty. But closer inspection reveals that this is not (necessarily) the case. The liberty which is entailed by the right is also entailed by the immunity, since

[26] See Raz 1984(*b*), 11–12.
[27] See Lyons 1970, 49 ff.; Hart 1982, 190 ff.

it is an immunity against legislative restriction of *that* liberty. The relationship between the liberty and the immunity is thus logical rather than normative: the liberty is part of the content of the immunity. Since every second-order normative position must take some first-order position as part of its content, there is a sense in which no second-order position is simple. But the logical complexity of second-order positions is irrelevant to the issue of whether at least some rights consist solely of such positions. Since the first-order liberty which is insulated from abridgement by a second-order immunity is already included in that immunity, we need not add it as part of a more complex package. Furthermore, since we can easily imagine the immunity also insulating the protective perimeter of the liberty against legislative interference, we need not add that either. Indeed, it may be that once we have fully spelled out the content of the immunity (i.e., the full set of first-order deontic positions which it insulates against change), nothing more will be needed in order for it to qualify as a genuine right. Certainly since the immunity is not a power it cannot require further protection by an immunity; nor, for the same reason, does it require a higher-order liberty to exercise it. An immunity thus appears not to need supplementation by any of the ingredients which we have found reason to add to bare liberties and powers. It seems reasonable to think that at least some immunities, all by themselves, are rights, and thus that at least some rights are just immunities.

Let us take stock of results so far. There seems little merit in the idea that a bare liberty (even a full liberty) can constitute a right since it can impose no normative constraints on others. There is even less merit in the idea that a bare power can constitute a right since it requires supplementation by a number of further ingredients. On the other hand, it seems reasonable to think that a bare immunity can constitute a right since it does impose a normative constraint (a disability) on others. However, we have no reason to suppose that all rights are just immunities, or even that all rights include immunities. We have thus not got far with defending the simple view that every right is identical with precisely one normative advantage. This is not surprising, however, for we have not yet considered the strongest version of this view. There still remains the possibility that (some or all) rights are just claims.

As we have seen, Hohfeld defended the view that rights 'in the strictest sense' are just claims. A similar view has been defended by

Feinberg.[28] Since claims impose deontic constraints on others it is easy to see why they have been singled out as capturing the insistent or peremptory quality of rights. Indeed, the plausibility in the idea that some rights are just immunities seems to derive from the fact that immunities, being the second-order analogues of claims, impose alethic constraints which are the analogues of deontic constraints. None the less, we already have decisive evidence against the (monistic) hypothesis that all rights are nothing but claims. If some rights are full liberties with a protective perimeter of duties, while others are full powers with a similar protective perimeter (plus additional ingredients), and still others are bare immunities, then some rights are not just claims. Indeed, surprisingly few rights consist of nothing but claims. The standard model for illustrating the claim–duty normative relation is that of two-party agreements in which the first party's claim against the second hooks up directly with the second party's duty to the first. But the making of agreements involves the power to impose duties on oneself (thereby conferring claims on the other party), while the terms of most agreements also involve the power to waive one's claims (thereby extinguishing the duties of the other party). Contractual rights are thus unintelligible without the notion of a power (and doubtless that of an immunity as well).

Are there any rights which are bare claims and nothing more? The right answer to this question depends on how we interpret an aspect of a claim which we have thus far left undefined: its directionality (and that of its correlative duty). What, then, does it mean for a claim to be held by one party *against* another, or a duty owed by this second party *to* the first? What makes a duty a relational duty? And which duties are relational? If we begin with this latter question then one simple answer is: all of them. Bentham sometimes seems to give this answer for the special case of legal duties. Thus, in a context in which by 'rights' he means what we mean by 'claims', he says: 'To create offences, is therefore to create obligations or forced services: to create obligations or forced services, is therefore to create *rights*.'[29] Since for Bentham every (coercive) law creates an offence, it follows that every such law also creates both a relational duty and its correlative claim. It is easy to

---

[28] Feinberg 1980, chs. 6 and 7. But note Feinberg's emphasis on the activity of claiming, which presumably requires the exercise of some powers.

[29] Bentham 1843, iii, 159.

imagine this account being generalized to cover the restrictive rules of any rule system. Thus far, however, it does not tell us what it is for a duty to be directional, and thus relational.

Bentham's account of the nature of relational duties turns on the notion of a beneficiary, and thus a benefit. On his view, I have a duty owed to you (and thus you have a claim against me) just in case the rule system imposes on me a duty from whose fulfilment you stand to benefit. This crucial notion of 'standing to benefit' is regrettably vague.[30] Bentham did not intend that the beneficiary of a duty actually benefit on every occasion of the performance of the duty. My duty therefore may be owed to you even if you are sometimes unexpectedly made worse off by my fulfilling it. What seems necessary is that fulfilment of the duty normally confer a benefit on you, or perhaps that it confer on you what would normally be regarded as a benefit, or perhaps that it be intended to benefit you, or whatever. Bentham is of little help here since he tends to speak of someone's being the party 'favoured by' the rules, or 'supposed' to benefit from them.[31] To the extent that these notions remain unclear, Bentham's analysis of the nature of relational duties also remains unclear.

However this may be, the implications which he extracted from his analysis, for the special case of a legal system, are tolerably clear. Bentham thought that there are two classes of duties which lack correlative claims: those whose infringement will cause no harm at all ('barren' duties) and those whose infringement will harm only the duty-bearer ('self-regarding' duties). Thus, despite initial appearances to the contrary, he did not hold the view that all duties are relational. The remaining duties Bentham divided into three categories. A duty is private if its infringement will harm some determinate individual(s) other than the duty-bearer, semi-public if its infringement will harm some indeterminate group of individuals within the community, and public if its infringement will harm the community as a whole though no one individual more than another. Thus the duty not to commit assault is private, the duty not to create excessive noise is semi-public, and the duty not to print counterfeit currency is public. Since all of these duties have (intended) beneficiaries Bentham treated all of them as relational.

It follows from Bentham's analysis that virtually all of the

---

[30] For some clarifications see Lyons 1969, 175 ff.; Hart 1982, 174 ff.
[31] See, for example, Bentham 1843, iii, 159.

(coercive or restrictive) rules of public law, plus those of private law, impose relational duties. The category of public duties is obviously the most problematic for this view. To whom is my duty not to print counterfeit currency owed? Is it owed to anyone? If we begin by treating the duties created by two-party agreements as the paradigms of relational duties, then my duty not to counterfeit will seem to have wandered a long way from that paradigm. Critics of the benefit analysis have standardly pointed to public duties as counter-examples, while even those friendly to the analysis have tended to disown Bentham's results in this area. Thus Austin treated public duties as non-relational and required that relational duties be owed to some determinate second party.[32] More recently, David Lyons has defended a qualified version of the benefit analysis according to which you are a beneficiary of my duty only if you benefit *directly*, or are *intended* to benefit, from my fulfilment of the duty.[33] From these stricter requirements for counting as a beneficiary Lyons draws the conclusion that public duties have no beneficiaries and are therefore non-relational.

The notions of benefiting directly, or of being intended to benefit, are still uncomfortably vague (and not obviously equivalent). However, Lyons's modification of the benefit account is doubtless an improvement. Its main virtue is that it directs our attention to the rationale behind the inclusion of a particular restrictive rule in a rule system. As Lyons says of those rules of criminal law which impose what Bentham called private duties:

Rules such as those forbidding murder and assault . . . can only be understood as requiring that we not harm or injure others in certain ways. The duties they impose patently require treating others in ways designed or intended to serve, secure, promote, or protect their interests. The rules define the classes of persons protected, and any member of such a class is a beneficiary in the qualified sense. He does not merely 'stand to benefit' by the performance of such a duty, nor does he merely 'stand to suffer' if the duty is breached, for his loss at the hands of the person with the duty would be directly relevant to the question whether the duty is breached. From the point of view of the rules and the duties they impose, such a person is neither a lucky bystander in the one case nor an unlucky bystander in the other. He is one who, according to the rules, is not to be harmed in such a way.[34]

---

[32] Austin 1885, 395.   [33] Lyons 1969, 176 ff.
[34] Ibid. 179; cf. Hart 1982, 175 ff.

The remaining defect in Lyons's account is that it does not tell us how to identify the rationale behind the inclusion of a duty-imposing rule in a rule system. However, some recent treatments of rights by Joseph Raz can be read as attempting to supply this needed further ingredient.[35] In the terms I have been using, Raz treats a duty as relational when the interest which it serves to protect or promote is the ground for the imposition of the duty. Thus whether a particular duty is relational is determined by the way in which the rule imposing it is justified within the system to which it belongs: a duty is owed to a particular class of individuals just in case the rule imposing it is justified by its role in serving some interest of those individuals. It is important to note that the justification which determines the relationality and non-relationality of a duty is relative to the rule system which imposes that duty. When Raz speaks of the justification of a rule he is not speaking of its moral justification (except in the special case of a moral rule). His version of a benefit account thus requires a background theory of normative justification, or practical reasoning. Raz's advance over Lyons lies in his attempt to supply such a theory.

We need not, however, pursue the complexities of a plausible version of the benefit account any further, since its general shape is clear enough. However it may be worked out in detail, it provides but one interpretation of the directionality of relational duties. Its chief competitor is an account which picks out as the object of a relational duty not the party whose interest is protected by the duty but the party who has effective control over the duty.[36] This account also treats two-party agreements as the paradigmatic setting for claim–duty relationships, but it focuses on the fact that in this setting a claim-holder is typically able to choose either to waive or renounce her claim (thus annulling the correlative duty borne by the other party) or to leave it in force. Moreover, where the agreement is legally enforceable she will also typically be able to choose either to seek its enforcement or to leave it unenforced, and where she secures a judgment in her favour she will be able to choose either to accept the resulting compensation or to waive it. It will be obvious from our previous discussion that these measures of control which the claim-holder enjoys over her claim (and thus over

---

[35] Raz 1984(a); 1984(b). But it should be emphasized that Raz's aim is to analyse the notion of a right, not that of a relational duty.

[36] See Hart 1982, ch. 7; Wellman 1985, 27 ff.

the relational duty) are so many Hohfeldian powers. Thus on the control account whether or not a duty is relational is determined by whether or not there is some party other than the duty-bearer who possesses this bundle of powers over the duty. H. L. A. Hart has provided the classic formulation for the special case of legal duties:

Instead of utilitarian notions of benefit or intended benefit we need, if we are to reproduce this distinctive concern for the individual, a different idea. The idea is that of one individual being given by the law exclusive control, more or less extensive, over another person's duty so that in the area of conduct covered by that duty the individual who has the [claim] is a small-scale sovereign to whom the duty is owed.[37]

Because Hart's version of the control analysis is developed for the case of relational duties in the law, it attaches considerable importance to powers of enforcement. If such powers are assumed to be necessary ingredients in the claim-holder's control over the correlative duty, then the account cannot easily be generalized to cover rule systems which lack institutions for applying and enforcing the rules.[38] In that case it would be quite unable to make sense of relational duties in non-institutional rule systems, such as those of informal associations. But the control account is much more plausible if it treats powers of enforcement as collateral to the claim-holder's one indispensable measure of control, namely her power to extinguish her claim and thus also its correlative duty. This power requires no institutional mechanism for enforcing duties and is a commonplace in both formal and informal rule systems.

We now have on hand two different accounts of the nature of relational duties. What practical difference does it make which we prefer? The special case of a legal system will serve to illustrate the areas of agreement and disagreement between the two accounts. They converge in treating most duties imposed under private law, especially under the law of contract, as relational. However, their convergence in this area results from the fact that in private law the party who stands to benefit from the fulfilment of a duty is normally also assigned powers over that duty. Wherever this regularity fails to hold, as it does in the case of third-party beneficiaries, then the two accounts will yield contradictory

[37] Hart 1982, 183.
[38] See Raz's critique of the 'institutional model' in 1984(*b*), sect. 1.

results.[39] They also converge in treating as relational some duties imposed on public officials by the provisions of a welfare system and in treating as non-relational many duties under the criminal law (Bentham's public duties).[40] But they diverge dramatically for the class of duties imposed by criminal legislation which Bentham called private duties. Because these duties are imposed in order to protect individuals from harm, the benefit account classifies them as relational. Because the beneficiaries of the duties have no power to extinguish them, and little or no control over their enforcement, the control account classifies them (in common with all duties under criminal law) as non-relational.[41] Thus whereas both analyses agree in recognizing the existence of some non-relational duties there are more such duties on the former analysis than on the latter.

Our interest, however, lies not in the classification of legal duties but in the question of whether some bare claims constitute rights. Recall that the right answer to that question depended on an account of the directionality of claims (and their correlative duties). What, then, are the implications of our two accounts for these issues? If we adopt the benefit account then it is not difficult to find claims unaccompanied by further normative advantages such as powers. Furthermore, since it is common to speak of the rules conferring these claims as protecting such rights as personal security, the benefit account will support the idea that some bare claims constitute rights and therefore also the idea that some rights are simple. On the other hand, if we adopt the control account the situation is somewhat more complicated. On this account I can have a claim only when I also have a power (or bundle of powers). But the power is not something I have in addition to my claim, since it is a necessary ingredient of it. Thus even on the control account we can make sense of having a bare claim, unaccompanied by any further normative advantages, and also of this claim counting on its own as a right. But this can be so, just as in the case of immunities, only because claims are themselves internally complex. Thus the control account lends no support to the idea that some rights are

---

[39] See Lyons 1969, 180 ff.; Hart 1982, 187.

[40] For the former, see Hart 1982, 185–6.

[41] The account in Wellman 1985, 27 ff., is an exception to this generalization. But Wellman focuses exclusively on powers of enforcement, thus on features peculiar to certain institutional rule systems. Cf. Raz 1984(*b*).

simple. Furthermore, if we accept the control account then we must also accept a complication in our two-level classification of Hohfeldian positions. If having a claim necessarily involves having a power, then claims are not only internally complex: they are also hybrids of first-order and second-order normative positions.

We began by entertaining the hypothesis that some or all rights might be simple, by virtue of consisting of a single Hohfeldian position. But we have found little support for this hypothesis. Since some rights are clearly bundles of positions, it cannot be true that all are simple. Furthermore, while some rights may indeed consist of a single position, even this fact may not entail that any rights are simple. The only two candidates capable of standing alone as rights are immunities and claims. By virtue of being second-order positions immunities must be internally complex. And if we accept the control account of the nature of claims then they too are internally complex. Thus only the benefit account is able to support the idea that any rights are normatively simple.

We are therefore inevitably drawn to hypotheses which treat most (or all) rights as packages of Hohfeldian normative advantages. But which packages of which advantages? Since liberties, claims, powers, and immunities admit of innumerable different combinations and permutations, it would be fruitless to embark on an exhaustive inventory of the possible structures of a right. Instead, we will turn to advantage our two models of the nature of claims by developing their analogues for rights. (Indeed, both of these models are standardly advanced as analyses of rights.) The analogue of the benefit account of claims is a conception of rights as protected interests.[42] Central to this conception is the idea of the right-holder as the beneficiary of a set of duties imposed on others, or as the one whose interest provides the justification for imposing such duties. The content of these duties may be either positive (to provide some good or service) or negative (not to harm), while their subjects may variously be individuals, or collectivities, or institutions. Different permutations of content and subject will doubtless yield different varieties of rights. If we build the benefit account of claims into this conception of rights as protected interests, then the beneficiary of

---

[42] This conception has been known in the literature as the interest or benefit theory of rights. Its recent defenders have included Lyons 1969; MacCormick 1977; 1982, ch. 8; Raz 1984(*a*); 1984(*b*).

this set of duties will necessarily be the holder of a claim. This claim may, but need not, be accompanied by further Hohfeldian advantages, such as the ability to waive or otherwise alienate it. Whether the right-holder enjoys any such further advantages or not, he has a right in virtue of having some interest protected by the duties borne by others. On this conception, therefore, a right may consist of nothing but a first-order Hohfeldian position.

The analogue of the control account of claims, by contrast, is a conception of rights as protected choices.[43] Central to this conception is the idea of the right-holder having the freedom to choose among a set of options, and of this freedom being protected by a set of duties imposed on others. The choice in question may be provided by a full liberty, in which case its protection will include claims of non-interference against others. But it may also take the simpler form of a claim, since (if we assume the control account) every claim necessarily involves the power either to demand performance by the duty-bearer or to waive it. In this latter case the choice will be protected by an immunity against the powers of others. Different permutations of liberties, claims, powers, and immunities will doubtless yield different varieties of rights. Furthermore, the levels of Hohfeldian positions involved in a right may be iterated indefinitely, first-order positions entailing second-order positions which in turn furnish the contents of third-order positions, and so on. On this conception, therefore, a right cannot consist of nothing but a first-order Hohfeldian position.

The distinction between these two conceptions of a right should not be confused with a classification of different kinds of rights. Different species of a common genus necessarily share a common conception of the nature of that genus and are thus not conceptual rivals. Different conceptions of a common concept, by contrast, offer rival interpretations of the nature of that concept.[44] Thus both the interest conception and the choice conception purport to tell us what rights really are. Better: each provides a model which will enable us to distinguish between standard or normal cases and non-standard or deviant cases. In establishing a paradigm or ideal each

---

[43] This conception has been known in the literature as the choice or will theory of rights. Its recent defenders have included Hart 1982, ch. 7; Wellman 1985, chs. 3 and 4.

[44] For the distinction between a concept and its conceptions see Rawls 1971, 5 and Dworkin 1977, 134.

conception is therefore itself normative, and consequently intolerant of the competing normative claims of the other.

Despite their differences, however, the two models are still conceptions of the same concept. They share a commitment to the root idea that the function of rights is to serve as one kind of constraint on the pursuit of social goals. Thus they share the conviction that real rights—standard, normal rights—must protect their holders by imposing normative constraints on others, and that these constraints must include duties borne by these others. Thus, whatever else rights may consist of, they must include claims. From this common starting-point the two conceptions diverge by offering rival interpretations of the point or function of rights. The interest conception treats rights as devices for promoting individual welfare. Thus the dominating image here is of the right-holder as the passive beneficiary of a network of protective and supportive duties shared by others, from which it follows that a being can be a right-holder only if it possesses interests. On the other hand, the choice conception treats rights as devices for promoting freedom or autonomy. Thus the dominating image here is of the right-holder as the active manager of a network of normative relations connecting her to others, from which it follows that a being can be a right-holder only if it possesses these managerial abilities. Since the two conceptions are not extensionally equivalent, there will be cases in which one will lead us to affirm the existence of a right which the other will lead us to deny. The crucial case which divides them is that of a duty borne by others whose function is to protect some interest of mine but over which I have no normative control; in such a case the interest conception will assign me a right while the choice conception will not. Since the interest conception can treat freedom or autonomy as a particular (higher-order) interest it will generally recognize as a right anything so recognized by its rival, whereas the reverse is not true. Furthermore, it will generally recognize any being capable of (the requisite sort of) autonomy as a being having interests, whereas the reverse is also not true. Thus the conception of rights as protected interests is likely to distribute rights more widely than the conception of rights as protected choices.

In particular, the distinction between the two conceptions should not be confused with a superficially similar classification of two kinds of rights. In the literature it has become common to

distinguish between liberty-rights and claim-rights.[45] The difference
between these two kinds of rights can best be delineated by utilizing
a distinction drawn by Carl Wellman between the core of a right
and its periphery.[46] The core of a right defines both its content
(what it is a right to) and its scope (who holds the right against
whom). A liberty-right is a right to do something or not as one
pleases; its core is therefore a full liberty.[47] A claim-right is a right
that something be done by another; its core is therefore a claim.
Since every determinate right must have some determinate content
and scope, every right must have a core. The periphery of a right is
then composed of any further ingredients which are added in order
to enhance or protect its core. In the case of a liberty-right these
further ingredients will include a protective perimeter of claims,
powers over both the core liberty and these protective claims,
immunities against the like powers of others, and so on. In the case
of a claim-right these further ingredients may (but need not) include
liberties which will enable the right-holder to utilize the service to
which she is entitled, powers to waive or otherwise alienate her
claim, immunities against the like powers of others, and so on.
Because the core of a liberty-right does not itself impose any
constraints on others, every liberty-right must also have a periphery.
Thus liberty-rights by their nature are structurally complex.
Because the core of a claim-right itself imposes a constraint on
others, some claim-rights may have no periphery. Nevertheless,
most claim-rights will also be structurally complex. The picture,
then, is of two different kinds of rights consisting of two different
bundles of Hohfeldian elements. Both the distinction between the
two kinds and the unity of each bundle is given by the core
elements. A right of either kind is not a random assortment of
Hohfeldian positions. Instead, it consists of nested layers of

---

[45] Liberty-rights are sometimes called active rights or rights to do, while claim-
rights are known variously as passive rights, rights of recipience, rights that
something be done, and rights to services. It should be noted that the distinction
which I draw here between liberty-rights and claim-rights does not coincide with the
distinction I drew in Sumner 1981 between liberty-rights and welfare-rights. Not all
claim-rights are welfare-rights.

[46] Wellman 1985, 81 ff.

[47] A liberty-right leaves room for what has been called the 'right to do wrong'
(Waldron 1981), as long as the wrong is not a violation of a duty (in the same rule
system).

components, each added in order to enhance the integrity or efficacy of its predecessors.[48]

Since liberty-rights presumably protect choices and since claim-rights can readily be construed as protecting interests, it is easy to confuse these two varieties of rights with the two conceptions of rights as protected choices and as protected interests. Both conceptions, however, are capable of accommodating both varieties. It is certainly true that the choice conception is most at home with liberty-rights, in virtue of the choice built into their core. But it readily recognizes claim-rights as standard cases of rights as long as the right-holder (in one way or another) has powers over the duties entailed by the core claim plus the (full) liberty to exercise these powers. Likewise, the interest conception is most at home with claim-rights, in virtue of the protective duties built into their core. But it readily recognizes liberty-rights as standard cases of rights as long as their periphery includes claims which protect the core liberty. Thus while the choice conception requires every right to contain a full liberty it need not insist that the liberty be located in the core, and while the interest conception requires every right to contain protective duties it need not insist that they be located in the core. In effect, the choice conception treats claim-rights as assuring higher-order choices while the interest conception treats liberty-rights as protecting one particular interest. Each has the conceptual resources necessary to recognize both kinds of rights as standard cases.

The adoption of a particular conception of a right will structure the way in which we view the nature and function of all rights. Since the two conceptions diverge over a crucial range of cases, it matters which we adopt. But how are we to decide between them? One obvious way of comparing them is in terms of their extension: what they include and exclude as genuine instances of rights. We might say, in general, that a conception of a concept is extensionally adequate when it includes every item which seems pre-analytically to be an instance of the concept and excludes every item which does not. It would then count in favour of a conception of a right that it draws the boundary between rights and other things in more or less

---

[48] The core/periphery distinction can be readily extended so as to define power-rights and immunity-rights as well; see Wellman 1985, 66 ff. I leave open the question whether these further categories are reducible to liberty-rights and claim-rights.

the right place, and against a conception that it draws it in the wrong place.

As soon as this test of adequacy is described, however, the problems of applying it become evident. The test is workable only if we can presuppose pre-analytic agreement on what counts as a right. But such agreement may be unavailable for the rights conferred by many rule systems. For one thing, it is very likely that our common notion of a right is loose, ambiguous, and vague. Where it is loose it may be used to embrace such a varied assortment of items that no internally coherent conception of a right can capture them all. Where the notion is ambiguous ideals of precision and clarity may require a conception to stipulate its 'strict' or 'proper' sense. And where it is vague a conception may have to draw an arbitrary boundary somewhere in its penumbra. All of this is to say that any interesting conception of a right will be not merely descriptive of our actual usage but also partially reconstructive or stipulative. But in tidying up our common notion a conception will inevitably expel from its extension some items which we pre-analytically recognize as rights (and perhaps also admit others which we do not recognize). Furthermore, there are bound to be disputed instances which some of us count as rights and others do not; in these cases no resolution of the dispute will satisfy everyone's intuitions. Finally, we should not expect that our conceptual intuitions, especially about a normative category as important as rights, will be genuinely pre-analytic. The purpose of a conception of a right is to provide a model which will illuminate the nature of rights, and thus a template for locating rights within a rule system. If our intuitions have already been corrupted by commitment to such a model then we cannot appeal to them as neutral arbiters among competing models.

For all of these reasons, therefore, we should not expect the test of extensional adequacy to settle the issue between our rival conceptions of a right. None the less, most of the debate between defenders of the two conceptions has focused on their extensional adequacy. Although this debate has largely been conducted for the special case of legal rights, the issues at stake can readily be generalized to other conventional rule systems. Thus proponents of the choice conception have claimed on its behalf that it alone can make sense of rights as capable of being *exercised*, whereas the interest conception entails that we could have rights which, because they are utterly beyond our normative control, are unexercisable.[49] But others have found nothing paradoxical in the notion that some rights are unexercisable.[50] Indeed, proponents of the benefit conception have turned this issue to their advantage by contending that it alone can make sense of the idea of inalienable rights, and also of the companion idea that beings incapable of exercising control over their normative relations—such as children or animals— might none the less be capable of having rights.[51] Proponents of the choice conception have also claimed on its behalf that it alone can explain why we feel uncomfortable with the notion of a mandatory right, since such 'rights' provide us with no discretion whether or not to engage in the activity in question.[52] But their opponents could reply that we speak of having the right to do what we have a duty to do only in those cases in which we regard the activity in question not merely as a burden but also as a benefit or privilege, thus lending support to the benefit conception. Finally, proponents of the choice conception have also claimed that it alone can explain why third-party beneficiaries lack rights.[53] But their opponents have countered with their own explanations of the normative standing of such beneficiaries.[54]

It is evident from these disputes that for each of the competing conceptions there are both easy cases of rights which it can readily accommodate and hard cases which it can accommodate only by dint of some delicate manœuvring. But this is precisely what we should expect from models which are partially stipulative. If a model, in its treatment of disputed cases, gives us good reasons for relocating the boundaries of our concept then we should be prepared to do so, at least as long as it is not mishandling any undisputed cases. The result of these extensional skirmishes is therefore essentially a stand-off. There is, however, a further test which has been implicitly or explicitly suggested by some commentators and which we could also treat as a standard of extensional adequacy.[55] It has been pointed out that rights are not merely outputs of a rule system, but instead play a justificatory role within the system. Once a right has been recognized within a system then it

[49] Hart 1982, 184. [50] See, for example, Montague 1980, 379 ff.
[51] MacCormick 1977; 1982, ch. 8.
[52] See Robinson *et al.* 1983, 268–9.
[53] Hart 1982, 187 ff.
[54] Lyons 1969, 180 ff.; MacCormick 1977, 208–9.
[55] See MacCormick 1977, 206; Raz 1984(*a*); 1984(*b*).

can be used to ground the introduction of rules which further enhance or protect the right. Thus a right is not a fixed or static bundle of Hohfeldian elements, but instead is capable of growth or development over time. Any conception of a right which fails to capture this dynamic aspect of rights should therefore be rejected as extensionally inadequate. Those who have emphasized this dynamic aspect have also tended to favour the conception of rights as protected interests. But they have not in fact argued that its rival is incompetent in this respect, nor is it easy to see how this could be argued. On the interest conception the protection of some interest can be used as the rationale for creating a unified bundle of Hohfeldian elements, but on the choice conception the protection of some liberty can be used in the same way. Thus both conceptions seem capable of explaining the internal coherence, the justificatory role, and the evolution of rights. A conception would fail this test only by treating rights as static assortments of Hohfeldian ingredients unified by no inner logic of their own. But there is no reason for either conception to treat rights in this way.

This last standard of extensional adequacy does, however, suggest a further and different means of comparing the merits of the two conceptions. A conception of a right is extensionally inadequate, we have said, if it fails to capture the ways in which rights function within the rule systems which create them. Suppose, however, we shift attention to the ways in which a conception of a right might function, or the work we might make it do. Then one conception might be functionally more adequate than another if it is better adapted to serving some important conceptual or theoretical purpose. Some of the debate in the literature concerning the two conceptions has been implicitly directed at their functional adequacy. Thus proponents of the choice conception have claimed that if rights are just beneficial duties then the notion of a right is redundant, since we can eliminate it in favour of talk about duties (and benefits).[56] Their opponents have replied by pointing to the justificatory role of interests and rights within the structure of a rule system, and by offering an account of how rights can be normatively prior to duties even though they are logically reducible to them.[57] Indeed, they might also (less constructively) have replied with a *tu quoque*: on the choice conception rights are reducible to

[56] Hart 1982, 181–2.
[57] MacCormick 1977, 199 ff.; Raz 1984(*a*); 1984(*b*); cf. Dworkin 1977, 171.

liberties plus claims plus powers plus immunities, in which case the notion of a right is presumably in principle eliminable in favour of talk about these more basic elements. Once again, the redundancy argument seems to cut neither way, since both conceptions are capable of explaining how the rationale for a right can lie in that which it protects (whether an interest or a liberty) and also how the right can in turn provide the rationale for further normative elements. Indeed, in explaining the dynamic function of rights within the rule system a conception is also explaining why the concept of a right is not redundant.

However, the redundancy argument points to a further, and more important, aspect of functional adequacy. Proponents of the choice conception have claimed on its behalf that it enables us to draw boundaries within a rule system in an illuminating way. Thus Hart, for example, has shown how the conception will group together the rules of private law (plus a few rules of public law) on the one hand and the rules of criminal law on the other. To the extent that this classification reveals patterns in the law which would otherwise be obscured, then the fact that it is generated by the choice conception should count in favour of that conception. However, since the interest conception could plausibly claim that it too draws lines in a theoretically fruitful way, this test is also unlikely to be decisive between the two conceptions. None the less, it is significant in reminding us that a conception of a right will play an important role in an explanatory or normative theory of law, politics, and morality. It remains possible, therefore, that one conception will enable us to construct our normative categories in a way which proves more fruitful or illuminating in this larger theoretical enterprise. Whether or not this is the case cannot be determined in advance of entering upon that enterprise. Meanwhile, therefore, we will have to tolerate a duality in our concept of a right.

3

Conventional Rights

THE strategy of the preceding chapter was to seek a coherent concept of a right by setting moral rights to one side and focusing instead on conventional rights. This order of inquiry seemed initially promising because the rights we enjoy by virtue of the conventional rule systems which apply to us appear to be accessible and determinate by comparison with the more abstract and mysterious moral rights. The pay-off from the strategy has, however, been unexpectedly abundant. We now have in stock two conceptions of a right: the models of protected interests and protected choices. Furthermore, we have as yet no decisive reason for preferring one of these models to the other. In the next chapter I will provide a theoretical argument in favour of the choice conception. Meanwhile, however, transacting the business of this chapter will be awkward if we must always keep both conceptions in play. Thus, for ease of exposition I will simply adopt the choice conception arbitrarily as my interim working model of a right. Nothing in the argument of this chapter would be substantially altered were we to work with the interest conception instead.

What then is our next step? Recall the nihilist's two broad challenges to moral rights: the conceptual claim that such rights are incoherent and the substantive claim that, even if coherent, there are no such things. Developing a coherent conception of a right has taken us some distance toward responding to the conceptual challenge. It has not taken us all the way, since the notion of a right might well be coherent while the notion of a *moral* right is not. And it has taken us no distance whatever towards responding to the substantive challenge, since the notion of a moral right might well be coherent while lacking any real instances. We are therefore still a long way from having existence conditions for moral rights.

There seem to be two directions open to us at this point. The first would involve completing our treatment of conventional rights by developing existence conditions for such rights. Our attention here would be focused on conventional rule systems, and especially on the conditions which such systems must satisfy in order to be 'in force', and our aim would be to uncover what it is for someone actually to have a conventional right. The second direction would involve completing our conceptual inquiry by developing a working model of a moral right. Our attention here would be focused on the several varieties of rights, and especially on the distinction between conventional and moral rights, and our aim would be to uncover what it is for a right to be a moral right. The mere depiction of these alternatives, however, settles their order of priority. Pursuing the second option would require an account of the distinction between conventional and moral rights. But thus far we have no such account; we have merely relied on an intuitive sense of the kinds of rules, and rights, which count as conventional. Only a set of existence conditions for conventional rights will tell us, *inter alia*, what it is for a right to be conventional. Since we know that moral rights, whatever they may be, are not *merely* conventional, this account seems likely to aid us in constructing a contrasting conception of a moral right. But in this case the first direction of inquiry logically precedes the second.

My concern in this chapter, therefore, will be to outline the conditions which must be satisfied in order for some particular subject to have some particular conventional right. We have a natural starting-point, furnished by the supposition that most, if not all, conventional rights are conferred or assigned by systems of conventional rules. The results of the preceding chapter, moreover, stipulate some conditions which must be satisfied if a rule system is to be capable of conferring rights. On the model of protected choices rights are unified bundles of first- and second-order Hohfeldian positions. It follows that a rule system can confer rights only if it contains rules which can confer liberties, claims, powers, and immunities—and thus also impose duties and disabilities. In order to be capable of defining these positions the rules of a system must be of the appropriate logical type; they must, that is, speak the appropriate (deontic or alethic) modal language. In this way the choice conception imposes semantic conditions on any rule system capable of generating rights.

A rule system may, however, satisfy these conditions while being merely ideal or fictitious or defunct. A system does not confer real rights on real subjects unless it exists and applies to those subjects.

Thus we now face a new set of questions: What is it for a conventional rule system to exist? What is it for such a system to apply to some set of subjects? What is it for rules to form a system? Do rules have to belong to a system in order to exist? What, indeed, is a rule? And what makes a rule, or a rule system, conventional? These are very large questions and providing full answers to them would be a complete inquiry in its own right. Although our ultimate interest lies not in conventional rights but in moral rights, we will not reach our destination if we side-step these questions entirely. Thus I shall aim in what follows not at providing full answers to them but at providing answers full enough to enable us to proceed. This will involve sometimes cutting corners and at other times relying on work done by others where it seems both promising and pertinent. I will not pretend that the story I am about to tell is original; with the exception of a few minor points it has all been said before. But it has not always, or often, been harnessed to the sort of inquiry which we have under way. Thus my immediate task is less one of creation or invention than one of adaptation and organization.

Just as conventional rights appear to be more tractable subjects than moral rights, some conventional rule systems appear to be more tractable subjects than others. The questions raised in the previous paragraph have been asked, and answered, more often for legal systems than for any other sort of rule system. (The same, of course, was true for the conceptual questions of the preceding chapter.) Legal systems furnish a natural point of departure both because they are denser and more fully articulated than most other rule systems and also because of the sophisticated literature available in analytic jurisprudence. I will attempt to capitalize on these advantages by first sketching existence conditions for the special case of legal rights; I will then try to extrapolate these results to other rule systems. We must remember, however, that legal systems are also distinctive among conventional rule systems in certain important respects; indeed, it is the features which are special to the law which explain the extraordinary amount of attention that has been devoted to it. We must therefore be careful not to build any of these features into our general category of conventional rights, nor to be overimpressed by ingredients which are prominent within legal systems but of little application elsewhere. In the end we want to identify both what is common to all conventional rights—what makes them all conventional—and also what distinguishes their many varieties.

3.1 LEGAL RIGHTS

We are seeking existence conditions for legal rules. In order to keep the inquiry under control I will confine attention to the rules which comprise municipal legal systems—the legal systems of sovereign jurisdictions—thereby side-stepping the special problems raised by international law. Recall that our notion of a rule, and thus also of a legal rule, is deliberately broad. Anything will count as a rule if it performs either a regulative or a facilitative function, thus if it determines either what those to whom it applies must (or must not) do or what they can (or cannot) do. Given this broad construal, we will count as legal rules all of the familiar contents of a legal system: statute law and case law, substantive rules and rules of procedure, criminal law and the traffic code, standards of tort liability, rules governing the enforcement of contracts, regulations administered by government agencies, the rules of the welfare system, tax regulations, family law, constitutional provisions concerning the separation of powers, the law of property, the provisions of a bill or charter of rights, and so on—all of the items, in short, with which lawyers typically concern themselves.[1]

A rule belonging to some municipal legal system may be said to exist when it is operative or in force within the jurisdiction of the system. Thus we are seeking the conditions which must be satisfied in order for a rule to be in force. Broadly speaking, two approaches seem to be open. What we may call the piecemeal approach involves seeking conditions which a rule may satisfy (or fail to satisfy) on its own, thus independently of its membership in a legal system. On this approach the existence conditions for a particular rule will consist entirely of circumstances pertaining to that rule, considered in isolation from the other rules of the system.[2] Indeed, on the piecemeal approach it is not necessary that a legal rule belong to any system in order to be in force.

One exemplification of the piecemeal approach is the view that a

---

[1] In particular, I will draw no distinction between legal rules and legal principles, such as that found in Dworkin 1977, ch. 2.

[2] This will be impossible for any rule which is internally related to some other rule(s) of the system; cf. Raz 1975, 112–13.

legal rule is in force just in case it is generally complied with. This view will obviously remain underdeveloped until we have in hand an explication of the notion of compliance with a rule. I will try to provide such an explication shortly. Meanwhile, we may get by with the following simplifying assumptions: the function of a legal rule is to regulate the activities of some set of agents, these agents comply with the rule only if they do what it requires (on the appropriate occasions), and the rule's efficacy is a (positive) function of the extent to which it is complied with by the agents to whom it applies. Then the view under consideration holds that a legal rule is in force just in case it is efficacious. Since the efficacy of a rule will plainly admit of degrees, further refinements of this view might tie the existence of a rule to its exceeding some stipulated threshold of efficacy or, alternatively, allow that the existence of a rule may itself be a matter of degree.

There is plainly something to the connection between a rule's existence and its efficacy. Efficacy might be part of the story for the existence of all conventional rules, and the whole story for the existence of some of them. But it cannot be the whole story for the existence of legal rules. The main problem with the piecemeal account is that a legal rule may be in force even though it is quite inefficacious. If the legislature passes a valid statute which the police and the courts attempt to enforce but which is widely ignored or flouted by those to whom it applies, it none the less remains the law of the land until it is repealed or otherwise superseded. A particular legal regulation may be complied with to a greater or lesser extent at different stages of its life history; through these fluctuations it does not pop into and out of force, nor is it in force to a greater or lesser extent. As long as it continues to be valid within the system to which it belongs, and as long as that system as a whole continues to be efficacious, then a particular rule may cease to be efficacious without thereby ceasing to be in force within the jurisdiction of the system. Being in force is not the same as being either enforced or enforceable; an inefficacious rule may still be a rule, but a nonexistent rule is no rule at all. A rule's existence cannot, therefore, be a direct function of its efficacy.

This latter problem with the piecemeal approach should incline us to a more global strategy of seeking existence conditions which a rule may satisfy (or fail to satisfy) only by virtue of its membership in a legal system. One instance of this approach is the view that a legal rule is in force just in case it is valid within a legal system which is itself generally efficacious. On this view we cannot understand what it is for a legal rule to exist until we understand what a legal system is, what it is for a rule to be valid within a legal system, and what it is for a legal system as a whole to be generally efficacious.

Since the existence conditions for a legal rule turn out on this approach to be very complicated indeed, all that I can undertake here is a sketch of how such a view might be elaborated. This sketch borrows heavily from the tradition of legal positivism, and especially from the shape which this tradition has taken in the work of H. L. A. Hart and, more recently, Joseph Raz.[3] We begin with an outline of the nature of a legal system. Such a system will consist in part of what Hart has called primary rules: rules which regulate the behaviour of private individuals. We may assume that these rules will perform their function either by imposing duties or by (implicitly or explicitly) conferring liberties. Their paradigms are the contents of the criminal law.

A set of primary rules appears to be one indispensable ingredient in a legal system. However, such a set is insufficient by itself to constitute a legal system, since it does not yet display the unity which is essential to such a system. That unity may consist partially in the fact that some rules of the system contain internal references to other rules of the system, and these references may enable us to make some sense of a hierarchy of rules. But the whole story of the unity of a legal system cannot be told in terms of primary rules alone. We need to add to the system rules whose function is to enable or facilitate by conferring legal powers and immunities. Some of these rules will, like primary rules, apply to private individuals, enabling them to make legally binding agreements, bequeath property, and so on. But others will confer powers whose effect is to make their holders public officials. Two varieties of these public powers are of special importance.[4] The first we may call legislative, since it consists of powers to create new rules of the system and to modify or annul existing rules. The second we may

---

[3] Hart 1961; 1982; 1983; Raz 1975; 1979; 1980; see also MacCormick 1978; 1981.
[4] See the distinction between rules of change and rules of adjudication in Hart 1961, 93–4, and the analogous distinction between norm-creating and norm-applying institutions in Raz 1975, 123 ff.

call adjudicative, since it consists of powers to apply the rules of the system to particular cases. Rules which confer legislative and adjudicative powers create special institutions (legislatures and courts) with the authority to perform special functions within the legal system, and they thereby make the members of those institutions public officials (legislators and judges) within the system. A legal system is thus, in this sense, an institutional rule system.[5]

The addition of rules which create and sustain authoritative institutions carries us some further distance toward explaining what makes a legal system a *system*. However, it does not carry us all the way; we also need to add what Hart has called a rule of recognition.[6] This rule lays down the fundamental criteria for the validity of a rule within the system. These criteria will consist of the various texts, traditions, institutions, or what not, which are to count as authoritative sources of the rules of the system. While these sources will doubtless vary somewhat from system to system, they will typically include both legislation (exercises of legislative powers) and adjudication (exercises of adjudicative powers). The rule of recognition enables us, directly or indirectly, to identify what is to count as a valid rule of the system, thereby providing the deep unity of the system. The validity of a rule is thus a systemic property: a rule is valid just in case it emanates, directly or indirectly, from something which the rule of recognition specifies as a source of valid rules.[7]

We have, therefore, a means of determining (in principle) whether a putative rule is valid within a given system, and thus the first ingredient in a global account of the existence conditions for legal rules. Before adding the further essential ingredient we should pause to note some complications. So far we have supposed that all the rules of the system either regulate conduct (by employing deontic modalities) or facilitate conduct (by employing alethic modalities).[8] In which of these categories does the rule of

---

[5] See Raz 1975, ch. 4, for an account of the nature of institutional rule systems.

[6] Hart 1961, 92–3, 97 ff. Hart appears to assume that a legal system can possess only one rule of recognition; for the contrary case see Raz 1975, 147, and 1979, 95–6. For simplicity the argument of this chapter will continue to assume a single rule of recognition. [7] See Raz 1979, 150 ff.

[8] We have also presupposed that some criteria are available for individuating rules. Depending on the criteria we select, it might be possible for one and the same rule to have multiple functions.

recognition belong? Hart appears to locate it in the second category.[9] However, the wiser counsel appears to be to regard it as a rule which regulates the conduct of adjudicative officials.[10] The rule of recognition for a legal system imposes upon these officials the duty of treating as valid all and only those rules which have issued from the sources specified as authoritative by the rule. The fundamental rule of the system—the rule in virtue of which it forms a *system*—thus has a regulative function. (Because it is addressed to the adjudicative officials of the system it is not, of course, a primary rule.) The fact that the rule of recognition is addressed to adjudicative officials also provides us with some reason for thinking that it is courts rather than legislatures whose existence is essential to the existence of a legal system.[11] A legal system which lacked any specialized legislative institutions is conceivable, though it would doubtless be inefficient; a system lacking adjudicative institutions is inconceivable.

A valid rule in a system is thus formally one which the courts have a duty to apply; materially it is one which has issued from some source specified as authoritative by the rule of recognition. This source-based test for the validity of a rule is the distinctive thesis of legal positivism.[12] One noteworthy implication of the test is that since it presupposes the rule of recognition it is incapable of validating that rule. Thus on this account the rule of recognition cannot itself be valid within its own system, and we will have to seek its existence conditions elsewhere. An account of these conditions is best postponed, however, until we consider the existence conditions for the system as a whole.

Thus far it has been convenient to assume that a rule's validity within a system is the same as its membership in that system. And it is certainly tempting to treat the test of validity also as a test of membership. But the temptation must be resisted.[13] In some legal systems courts may be obligated to apply rules which are not

[9] In Hart 1961, 78–9, he identifies secondary rules and power-conferring rules. The rule of recognition is, of course, a secondary rule.

[10] This option is urged in Raz 1979, 92–3 and 1980, 199–200, and in MacCormick 1981, 21.

[11] See Raz 1975, 132 ff.; 1979, 87–8; 1980, 191. For simplicity the argument of this chapter will continue to assume that legal systems contain both sorts of institution.

[12] I side here with Raz 1979, 47, against Lyons 1977(*b*), 423–5, on the issue of whether positivists can admit non-source-based criteria of validity.

[13] As is shown in Raz 1975, 152–4; 1979, 149.

themselves rules of the system within which the courts have jurisdiction. Under the rules of private international law, for example, the courts of one jurisdiction may be obligated to apply to a particular case the law of some foreign jurisdiction. In such cases although this law may not count as part of the domestic jurisdiction it is valid within that jurisdiction, since the courts are duty-bound to apply it. Validity within a legal system is necessary for membership in that system, but it is not sufficient. In order to construct conditions of membership which would be both necessary and sufficient we would require criteria for distinguishing different municipal legal systems (thus drawing the boundary between the domestic system and all foreign jurisdictions) as well as criteria for distinguishing legal from non-legal systems (thus drawing the boundary between the domestic legal system and other domestic rule systems whose rules courts might be obliged to apply). We will need to give some attention to this latter distinction when we begin to generalize the notion of a legal system and a legal rule. But for the moment there is no need to concern ourselves about the discrepancy between criteria of validity and criteria of membership. A rule is in force in a system if it is valid within that system, regardless of whether it is also a member of the system. Existence conditions for legal rules, and thus for legal rights, require only criteria of validity.

The validity of a legal rule is one ingredient of its existence. But a legal rule is in force only when it is valid within some legal system which is itself in force. Rules which are valid within ideal, fictitious, or defunct legal systems are themselves merely ideal, fictitious, or defunct. We therefore still need an account of what it is for a legal system as a whole to exist or to be in force. It is at this level of the entire system, rather than that of a rule within the system, that the test of efficacy applies. It is not necessary to the existence of a legal rule that it be efficacious, only that it be valid. But it is necessary to the existence of a legal system as a whole, and thus also to the existence of the rules which are valid within it, that it be efficacious.

The efficacy of an entire legal system is inevitably a complex matter. One complication, of course, is that on any reasonable account efficacy is bound to be a matter of degree. Thus whether a legal system is in force over some territory will not be an all-or-nothing affair; the system as a whole may be in force to a greater or lesser extent, and different parts of it may be in force to different extents. That it throws up this complication is not, however, a defect in the present account; indeed, since being in force does seem to admit of degrees it is a virtue in the account that it can support and explain this fact. A more serious complication results from the fact that the test of efficacy may vary for different components of the system. Thus Hart proposed one test for primary rules and a different one for secondary rules (the rules regulating the activities of public officials, and above all the rule of recognition).[14] Hart's two tests will be better understood if we distinguish three different ways in which a rule imposing a duty on some set of agents may succeed or fail in being efficacious. These three ways will be presented in order of increasing strength; that is, later tests entail earlier ones but are not entailed by them.

The first test I shall call *conformity*. I conform to a rule which imposes a duty on me just in case I do what the rule requires (on the appropriate occasions). Thus I conform to a rule prohibiting murder just in case I do not commit murder. Conformity to a rule is entirely a matter of overt behaviour. It is sufficient for conformity that I do what the rule directs; my reason for so acting is utterly irrelevant. That reason might be a conviction of the legitimacy of the rule or a desire to avoid the disagreeable consequences of violating it, but it also might have nothing whatever to do with the rule. Indeed, in order to conform to a rule I need not even know that it exists or applies to me. That a rule be generally conformed to—that nonconformity not be too widespread—is the minimal condition of its efficacy. Where a rule is generally conformed to for reasons having nothing to do with the rule we might, for obvious reasons, hesitate to say that it is efficacious. But a rule which is generally not conformed to is plainly inefficacious.

A stronger test of efficacy will require not just conformity but what I shall call *compliance*. I comply with a rule which imposes a duty on me just in case I conform to the rule at least in part *because of my awareness of it*. Compliance therefore presupposes conformity but places limits on the reasons for conformity. In order to comply with a rule I must know that the rule applies to me and this knowledge must play some motivational role in my conforming to the rule. Thus it must be true that ignorance of the rule might have made a difference to my conduct; I might in that case have failed to

[14] Hart 1961, 109–14.

conform to it. (It is not necessary that I *would* in that case have failed to conform, since compliance requires only that awareness of the rule play some motivational role for me; had I been ignorant of the rule I might still have had other reasons for conforming to it.) While the compliance test requires that my conformity to a rule be influenced in one way or another by the existence of the rule, it does not stipulate any particular form of influence. Thus I may conform to the rule because of a conviction of its legitimacy, but I may equally do so out of the fear of being caught. While my conduct has been motivated at least in part by my awareness of the rule, my motive need not have been more than ordinary prudence.[15] That a rule be generally complied with—that non-compliance not be too widespread—is the more-than-minimal condition of its efficacy. Notice that because compliance requires conformity for the right (sort of) reason, non-compliance can involve either nonconformity or conformity for the wrong reason.

A still stronger test of efficacy will require not just compliance but what I shall call *acceptance*. I accept a rule which imposes a duty on me just in case I comply with the rule at least in part *because of my endorsement of it.*[16] Acceptance therefore presupposes compliance (and therefore conformity) but places further limits on the reasons for conformity. In order to accept a rule I must in one way or another endorse it and this endorsement of it must play some motivational role in my conforming to the rule. Thus it must be true that failure to endorse the rule might have made a difference to my conduct; I might in that case have failed to conform to it (though once again I also might not). My acceptance of a rule requires that I regard the rule as more than just an externally imposed constraint on my conduct. I must be favourably disposed toward the rule, must regard it as in some respect or other reasonable or legitimate, and therefore must regard non-conformity as a ground for criticism or blame. In short, I must use conformity to the rule as a standard for judging or evaluating the conduct of those to whom the rule applies; I must, as we say, have internalized

---

[15] My standard of compliance with a rule is the same as Raz's standard of acceptance of a rule; see Raz 1980, 235. I introduce a stronger standard of acceptance below.

[16] On this account accepting a rule entails conforming with it. In fact the connection between acceptance and compliance must be looser, to allow for the fact that agents may sometimes fail to follow the rules which they accept. No harm is done to the present discussion, however, by assuming the tighter connection.

it.[17] That a rule be generally accepted—that rejection not be too widespread—is the maximal condition of its efficacy.

Hart's two tests for the efficacy of a legal system may now be easily stated. For the system's primary rules it is both necessary and sufficient that they be generally complied with.[18] Thus private citizens must by and large conform to the rules applying to them and do so at least in part because the rules apply to them. It is clear that their conformity to the rules is by itself insufficient, since the members of a society may, entirely unwittingly, conform to the rules of a system which is merely ideal, fictitious, or defunct. On the other hand, it is also clear that their acceptance of the rules is unnecessary. It is true that in any well-ordered legal system we should expect the compliance of members of the general populace to be motivated at least in part by their endorsement of the system as a whole. Thus in a well-ordered legal system we should expect most citizens to accept the rules of the system in addition to complying with them. But general acceptance by the populace, however desirable it might be, is not necessary in order for a legal system to be efficacious. It is sufficient that the rules applying to private citizens be generally complied with by those citizens, however grudgingly or reluctantly.

The situation is quite different for the secondary rules applying to the officials of the system, and above all for the rule of recognition. Here it is necessary that the rules be not merely generally complied with but also generally accepted. This stronger demand stems from the central and unifying role played by the officials of a legal system, especially by those who exercise adjudicative powers within it. Recall that the unity of a legal system—the fact that its various components form a *system*—resides largely in the test of validity contained in its rule of recognition. Thus a legal system can be in force only if its rule of recognition is in force; the existence of the latter is crucial to the existence of the former. As we saw earlier, the rule of recognition is unique among the components of a legal system by virtue of the fact that its validity cannot be established by its own criteria. Its existence can therefore consist only in its efficacy; it is the one rule of the system of which this is true. As we

---

[17] See Hart 1961, 55–6, 86–8, on the 'internal aspect' of rules.

[18] My notion of compliance is the same as Hart's notion of obedience. A full account of the efficacy of primary rules would need to expand the test of compliance so as to include rules which confer private powers; cf. Raz 1980, 203–5.

also saw earlier, the rule of recognition imposes upon adjudicative officials the duty to apply just those rules which pass its source-based test of validity. Its efficacy, therefore, depends on the practice of those officials. As a minimum this requires that they comply with it, thus that they recognize as valid all and only those rules which emanate from the authorized sources. But the nature of their adjudicative role requires additionally that they accept it. The practice of adjudication, especially at the appellate level, requires that decisions be supported by means of legal arguments. This practice in turn requires the shared acknowledgement by adjudicative officials of standards of correct legal argument. Since one way in which a decision can be incorrect is for it to ignore relevant aspects of the existing law, some of these standards will be provided by the rule of recognition which stipulates what is to count as the existing law. This means that their institutional role requires adjudicative officials to treat the rule of recognition as a standard of correct decision-making, both in their own case and in that of other such officials. It thus requires them to treat deviations from the rule as grounds for criticism or censure. But to treat a rule as a standard of correct behaviour within its domain is to endorse it and thus to accept it.[19] The special institutional role of adjudicative officials therefore requires general acceptance on their part of the rule of recognition. A rule of recognition is thus in force only when it is generally accepted by the officials to whom it is directed.

We now have a complete, though also sketchy, account of the existence conditions for legal rules. When we add this account to a conception of a right the result is a set of existence conditions for legal rights. If we continue to assume the model of rights as protected choices, then those existence conditions will run somewhat as follows. I possess a right under a given legal system when rules valid within that system confer upon me a suitable bundle of liberties, claims, powers, and immunities, and when the system as a whole is efficacious. What counts as a suitable bundle of these positions will be determined by what is necessary in order to assign me effective control over some specified domain, and above all by the functional relationship between the elements in the core of the right and those in its periphery. The rules which confer these various positions apply to me when I am included within their

---

[19] Endorsement need not involve moral approval; see Raz 1975, 147–8; 1979, 154–5; Hart 1982, 153–61, 264–8.

scope, thus also within the jurisdiction of the system as a whole. The rules are valid within their system when they have issued from sources of valid law specified by the system's rule of recognition. And the system as a whole is efficacious when its rules applying to private individuals are (at least) generally complied with and its rules applying to public officials (especially the rule of recognition itself) are generally accepted.

It is an important feature of this account that the existence conditions for legal rights are supplied by two different sorts of social practice. The content and scope of particular rules of the system, including those which (individually or collectively) confer rights, are determined by the decisions of legislative and adjudicative institutions. Thus legal rights exist only where they have been accorded official recognition within a legal system. The existence of a legal system is in turn determined by the general social practices of complying with and accepting its rules. Both the foreground practices which create and shape the particular rules of the system and the background practices which sustain the system as a whole are social phenomena whose existence can, in principle, be ascertained by ordinary empirical methods. The existence conditions for legal rights thus consist of layers of social practices; in this sense, legal rights are themselves social facts.

The question whether some particular legal right exists is thus an empirical one. To say this is not to say that the right answer to the question will always be clear, nor even that there will always be a right answer. The social practices on which the existence of legal rights depends will always be complex, and may sometimes also be conflicting or indeterminate. Thus while we should expect clear cases of undisputed legal rights we should also expect unclear cases. Within a settled and stable legal system most cases will be clear cases. When we are dealing with such a system we may assume that the background conditions for the existence of the system as a whole are satisfied, and that it is clear which rules are valid rules of the system. In that case disputes about legal rights will generally be disputes about the interpretation of the rules. These more narrowly legal disputes may themselves admit of no agreed, or even no uniquely correct, resolution. Some cases may therefore be intractably unclear, but these cases will occur in the gaps between the larger areas of settled law.

Legal rights are the products of social practices, both institutional and non-institutional. These practices endow legal rights with three features to which it is worth drawing attention. The first is their artificiality. To say that legal rights are artefacts is simply to point to the fact that they are created, sustained, and extinguished by the decisions of human agents. Where a given right exists it does so in virtue of an elaborate pattern of interlocking and mutually reinforcing decisions: on the one hand, decisions by legislative officials to create rules and by adjudicative officials to interpret and apply them and, on the other hand, decisions by private citizens to comply with primary rules and by public officials to accept secondary rules. If this pattern of decisions alters in certain ways then the rights sustained by it will also alter. Thus while the existence of a legal right will seldom be under the control of any agent acting individually (though this may be the case), it will be under the control of many agents acting collectively. The collective decisions of private and public agents *make it true* that a right exists; different collective decisions would make it true that the right does not exist. Legal rights are thus social creations.

The second feature of legal rights is their content-independence.[20] This feature follows from their artificiality. Because the existence conditions for a right consist solely of various collective decisions, those conditions can in principle be satisfied whatever the content of the right. That a given legal right exists in a given jurisdiction is entirely a matter of its being recognized in the appropriate way within that jurisdiction. The existence conditions for legal rights are in this respect formal rather than material. It follows that the contents of legal rights may vary widely from jurisdiction to jurisdiction (which, of course, they do). The practices of one society may make it true that a particular right exists in that society, while the different practices of another society may equally make it true that the same right does not exist in that society.

This content-independence of legal rights must not, however, be overstated. It is true that the existence of a legal right is, oversimplifying somewhat, a matter of its recognition within a legal system rather than of its content. It is thus true that, logically speaking, the only restrictions on the content of legal rights are conceptual ones. However, making even very weak assumptions about the aims of a legal system will impose restrictions on the rules

[20] See Hart 1958; 1982, 254–5.

of the system, and thus further restrictions on the content of the rights which the system recognizes. Hart has argued that if we assume that a legal system has as its aim the continued survival of its society then, given some universal features of our nature and our environment, there is a core content—what Hart called the minimum content of natural law—which will be the common property of all legal systems.[21] To this extent there are natural limits to the content of legal rules, and thus also to the content of legal rights, though it will still of course be true that many different, and mutually incompatible, sets of rules (and rights) will all fall within these natural limits.

The third feature of legal rights is that the background social practices which create and sustain them display a considerable degree of co-ordination. As we have seen, the ingredients of these practices are countless individual decisions of compliance and acceptance, both by private citizens and by public officials. These decisions are standardly made not in isolation but in the light of the like decisions of others, and thus in the light of the existence of the general practices. The individual decisions are thus both interdependent and, to some extent, also mutually reinforcing. Private citizens will normally be aware that they and their fellows are subject to a common legal system, and their decisions to comply with the rules of that system will normally be influenced by their expectation that others will do so as well. Each agent will therefore be motivated to comply partly by the confidence that the rules of the system will generally be complied with. This co-ordination of individual decisions, which is likely to hold in the case of private citizens, cannot fail to hold for the officials of the system. In order for a legal system to be in force its adjudicative officials must generally accept its secondary rules, and especially its rule of recognition. But accepting this rule involves using it as a general standard for evaluating the conduct of adjudicative officials. The role of such officials therefore requires knowledge of the institutional setting of their activities, and of the standards applying to those activities. But this means that acceptance by each official of the rules of the game will reflect awareness of the acceptance of the same rules by other officials of the system, and thus awareness of the existence of the general practice of rule-governed adjudication.

[21] Hart 1961, 189–95.

Where the individual decisions which comprise a social practice display this kind of co-ordination I will say that the practice as a whole is a *convention*.[22] The background social practices which create and sustain a legal system, and thus also the rights conferred by the rules of the system, are conventions. This is the sense in which legal rules are (one variety of) conventional rules, and therefore also the sense in which legal rights are (one variety of) conventional rights.

3.2 INSTITUTIONAL RIGHTS

The account which I have given of the existence conditions for legal rules, and thus also for the rights conferred by those rules, is one version of legal positivism. I am conscious of the fact that this account has been little more than an outline which stands in need of both further elaboration and further defence. There is no general agreement in the jurisprudential literature that any version of positivism is correct. Furthermore, even if some version of it is correct, it is very likely that the version I have sketched would turn out on closer examination to be mistaken in some of its particulars. None the less, I intend in what follows to assume that this version is correct in all essential respects. My reasons for proceeding on the basis of this rather peremptory assumption are entirely pragmatic. Although we need some account of the existence conditions for legal rights in order to anchor a generalized account for conventional rights, we do not need to work out all of its details. Thus it will be sufficient for present purposes to indicate the overall shape of what seems, initially at least, a plausible story. Given the size and richness of the jurisprudential literature, it would divert us much too far from our main item of business to try to construct a full account or to give it a full defence. Furthermore, much of that work has been done by others, chiefly by Hart and Raz. Enough has been said here to stand as a presumptive case in favour of a positivist approach to legal rules. Therefore, until that case has been rebutted

---

[22] My notion of a convention is somewhat weaker than that in Lewis 1969. The practices which I treat as conventional satisfy some, but not all, of Lewis's conditions for the existence of a convention. In particular, they fail the condition that each participant prefers that everyone conforms to the rules rather than that everyone else conforms while he/she does not. This is because most legal rules do not have the function of solving co-ordination problems.

I will continue to assume that the version of positivism offered in the preceding section is substantially correct.

We are now in a position to begin expanding our account of legal rights. My eventual aim is to show how legal rights are the most highly developed species of conventional rights. In order to achieve this aim it is necessary to isolate those features which distinguish them from the other members of the family. It will be convenient to undertake this task in two stages. In the first stage we will confine ourselves to rights conferred within institutional rule systems. At the completion of this stage we will have in hand an account of institutional rights which will be broader than the account of legal rights (since legal systems are but one variety of institutional rule systems) but narrower than an account of conventional rights (since institutional rule systems are in turn but one variety of conventional rule systems). The second stage will then involve expanding the account of institutional rights into an account of conventional rights, by showing that non-institutional rule systems are also capable of conferring rights.

A legal system is necessarily an institutional rule system. There are two respects in which this is so, of which one is more fundamental than the other. The first is that a (municipal) legal system is necessarily the rule system of some institution—namely, some sovereign state. The second is that a legal system necessarily contains specialized legislative and adjudicative institutions whose function is to create and apply the rules of the system.[23] I will treat the possession of such institutions as the distinguishing feature of an institutional rule system. For simplicity, I will also speak of a rule system as belonging to an institution, or as being the rule system of an institution, when and only when it possesses this feature. An institutional rule system is thus by definition also the rule system of some institution.

A legal system is necessarily an institutional system, but it is not the only sort of institutional system. The rule systems of many organizations besides the state will satisfy the requirements of an institutional system. Three such requirements are basic: (1) a set of primary rules applying to those subject to the authority of the organization, (2) a set of secondary rules establishing institutions

---

[23] As noted earlier, it is technically possible for a legal system to lack a specialized legislative body. However, for convenience I continue to assume the indispensability of both sorts of institution.

within the organization empowered to create and/or apply the rules of the system, and (3) a rule of recognition containing criteria of validity for the rules of the system. These requirements will be met by the rule systems of most organizations with a formal decision-making structure, thus by the rule systems of most corporations, businesses, trade unions, churches, schools, political parties, charitable agencies, professional organizations, military forces, clubs, governing bodies of sports and games, mafia families, and the like. The constitution of a political party, for example, may empower one specialized body (a convention of elected delegates) to create or modify its rules while empowering a distinct specialized body (the party executive) to apply these rules in particular cases. The party constitution may also, implicitly or explicitly, require the executive to apply all and only those rules which have been duly adopted at a properly constituted party convention. The rule systems of a great many organizations will therefore display the structural features which Hart pointed to as the mark of a legal, as opposed to a pre-legal, system.

What then are the features which distinguish legal from non-legal systems? In order to answer this question it will be convenient to assume that we can individuate municipal legal systems and arrange them in a one-to-one correlation with separate sovereign states. Our problem then is to distinguish the legal system of a particular state from the various non-legal institutional rule systems existing (wholly or partly) within that state.[24] The most obvious distinguishing feature concerns the range of activities which a rule system sets out to regulate. Most institutions exist for some special and limited purpose: to provide rail transportation, to organize track meets, to liberate political prisoners, and so on. Only certain activities therefore fall within the jurisdiction of such institutions; they claim the authority not to regulate all aspects of the lives of those subject to their rules but only those aspects connected with the special purpose or function of the institition. Thus the possible contents of the rule systems of most non-legal institutions are limited. The state, by contrast, does not limit itself to any special purpose or to any definite set of special purposes. It claims, in principle at least, the authority to regulate any activity on the part of those who are subject to its jurisdiction; no aspect of its subjects'

---

[24] The account which follows owes much to Raz 1975, 149 ff.

lives is guaranteed immunity to such regulation. Of course, no legal system in fact attempts to regulate every aspect of its subjects' lives and many systems contain restrictions, such as those embodied in a bill of rights, on the exercise of their authority. But where these restrictions exist they are of necessity self-imposed; legal systems which contain them thus also claim the authority to remove them. The authority claimed by legal systems is in this sense *comprehensive*.

The comprehensiveness of a legal system determines the relationship between the state and the various non-legal institutions operating (wholly or partly) within its jurisdiction. If the legal system claims the authority to regulate all of its subjects' activities then it must claim the authority to regulate the activities of non-legal institutions operating within its jurisdiction. The rule systems of such institutions will be formally autonomous if they are not parts of the legal system, thus if they possess autonomous rules of recognition. However, the range of activities regulated by these rule systems, the manner in which they regulate them, the procedures by which the rules are applied to particular cases, the range of penalties imposed in order to enforce the rules—all these aspects of the rule systems may be regulated by law. The legal constraints imposed on these organizations by the state will of course vary from case to case, and some may escape effective regulation altogether. As with comprehensiveness, the question is not what the legal system in fact does but what it claims the authority to do. The authority claimed by legal systems is in this sense *supreme*.

Comprehensiveness and supremacy lead us to a third distinguishing feature of legal systems. It is true of most non-legal institutions that subjection to their jurisdiction is, at least to some considerable degree, voluntary. The rules of many organizations apply only to members of the organization, membership itself being voluntary. In other cases one may become subject to the rules of the organization, or compelled to join it, as a condition of engaging in some specific activity. In all of these cases subjection to the authority of the institution is conditional upon pursuit of the ends which the institution promotes or the activities which it governs. Thus the authority of the institution can be evaded by foregoing those ends or activities. Because the state does not limit its jurisdiction to some special purpose, and thus does not limit itself to some special range of activities, its authority cannot be evaded in the same manner. The authority claimed by legal systems is thus also *compulsory*.

The compulsoriness of legal jurisdiction emerges most clearly when the geographical nature of the state is kept in mind. Unlike most other institutions, the boundaries of a state, and thus also the limits of its jurisdiction, are spatial. Within its boundaries the state claims the authority to regulate the activities of members (citizens) and non-members (aliens) alike.[25] The only condition one need satisfy in order to be subject to the authority of a legal system is to be located within the spatial boundaries of its jurisdiction; subjection to its authority can therefore be evaded only by removing oneself beyond those boundaries. The authority claimed by legal systems is thus also *territorial*.

The final feature which distinguishes legal from non-legal systems concerns their use of sanctions. Resort to sanctions in order to enforce rules is not, of course, a practice peculiar to legal systems; non-legal institutions may likewise impose penalties on those who infringe their rules. Nor is the employment of sanctions, strictly speaking, a necessary feature of a legal system.[26] But sanctions are in fact universal in existing legal systems, and it is typical of a legal system to reserve for its exclusive use the more severe types of penalties. Non-legal institutions may fine offenders, or withdraw their privileges, or suspend them from activities, or expel them from the organization. But they typically may not imprison them, or transport them, or mutilate them, or kill them—because they are legally forbidden to do so. This distinction between legal and non-legal institutions is thus a by-product of the supremacy claimed within its jurisdiction by the legal system. None the less, it is a sufficiently striking and universal feature of legal systems as to warrant separate mention. Let us say, then, that the authority claimed by legal systems includes *a monopoly on the use of the more serious forms of force and violence*.

If we gather these features together, legal systems may be distinguished from non-legal institutional rule systems by their claim of comprehensive, supreme, and compulsory authority within territorial boundaries, including the authority to regulate the use of force and violence by other institutions. These various features appear to be individually necessary; a rule system which failed to satisfy any one of them, at least to some degree, would not be a

[25] This claim will be limited in practice by such devices as the recognition of diplomatic immunity.
[26] Raz 1975, 157 ff.

legal system. They are also jointly sufficient. The rule systems of some non-legal institutions, such as churches or mafia families or paramilitary organizations, may display some of these features, such as the claim to comprehensive and/or supreme authority. But they will not display all of them, for if they did then they would be claiming the status of rival or parallel legal systems.[27] A legal system is thus one special case of an institutional rule system. The properties which distinguish legal from non-legal systems also, however, serve to explain the greater prominence of the former. A particular non-legal institution may impinge upon my life if it promotes some end which I wish to pursue or regulates some activity in which I wish to participate, or if its activities interfere with mine. But different non-legal institutions are likely to be important to different degrees for different individuals, and for the same individuals at different times. All of us, however, share a common stake in an institution which claims the authority to regulate all aspects of our lives as long as we remain within its boundaries and which threatens us with severe penalties if we violate its rules, especially if it has the wherewithal to make good both the claim and the threats. From a purely practical point of view, it is understandable that the state should nowadays be the foreground institution for most of us most of the time, and the legal system the most prominent rule system to which we are subject. It is small wonder, then, that the legal system should have come to serve as our model of an institutional rule system, and legal rights as our models of institutional rights.

The rule systems of non-legal institutions are, however, quite capable of conferring rights on those to whom they apply. The existence conditions for rights in such systems are easily extrapolated from the existence conditions for legal rights. I possess a right under a given institutional rule system when rules which are valid within that system confer upon me the appropriate bundle of liberties, claims, powers, and immunities, and when the system as a whole is efficacious. The rules which confer these Hohfeldian positions apply to me when I fall within their scope, which requires in turn that I fall within the jurisdiction of the system as a whole. The requisite rules are valid within the system when they issue from one of the sources of valid rules specified in the system's rule of

---

[27] Claiming that status is not, of course, the same as achieving it. Efficacy remains the test of achievement.

recognition. And the system as a whole is efficacious when the rules applying to private individuals are (at least) generally complied with and the rules applying to officials of the system are generally accepted. All of the principal ingredients of our account of legal rights apply straightforwardly to other institutional rights.

However, two observations are in order about the existence conditions for non-legal institutional rule systems. First, because subjection to the authority of these systems is generally more voluntary than subjection to legal authority, we should ordinarily expect more than mere compliance on the part of private individuals with those rules applying to them. Acceptance of the legitimacy of the institution, and of at least most of its rules, should be more common than is usually the case for a legal system. Second, because non-legal systems are themselves subject to the authority of the state, their existence may itself be partly a matter of law. Where the state has authorized some non-legal organization to regulate some domain of activities we will have a condition for the existence of the organization which is independent of the practices of those who are subject to it. Examples which come readily to mind include the governing bodies of self-regulating professions, such as law and medicine. However, in such cases we should not make the mistake of thinking that this further legal existence condition is necessarily decisive by itself. Resistance to the authority of the organization on the part of those whom it purports to govern may deprive it of all but a paper existence. Background social practices of general compliance and acceptance remain in all cases factors in the efficacy, and thus in the existence, of an institutional rule system. Furthermore, these practices will display the features of interdependence and co-ordination which were evident in the case of a legal system. Institutional rules of all sorts are thus conventional rules, and institutional rights of all sorts are conventional rights.

While it is usually clear whether a particular legal system is in force over a particular territory—though it may be unclear during times of rebellion or civil war—there may more frequently be room for doubt about the efficacy, and thus the existence, of other institutional rule systems. None the less, here too uncertainty is the exception rather than the rule. When we ponder the many non-legal institutions which play important roles in our lives—the organizations which employ us, educate us, advance our occupational interests, do business with us, protect our neighbourhoods, care for our health, promote our leisure activities, and so on—we are generally in little doubt about either their efficacy as a whole or the validity of their rules. Thus in these cases, as in the case of the legal system, questions about our rights are first and foremost questions about the contents of the rules which apply to us. A conception of a right imposes some very general demands on the rules of a system if it is to be capable of conferring rights. Under the conception of rights as protected choices, a rule system must contain rules of two distinct logical types: those which regulate action by stipulating what may or may not be done, and those which facilitate action by stipulating what can or cannot be done. The former confer liberties and claims and impose duties while the latter confer powers and immunities and impose disabilities. But containing rules which speak the appropriate modal languages is not enough by itself. A right is not just a random assortment of Hohfeldian positions but a coherent whole whose core is enhanced and protected by its periphery. A rule system capable of conferring rights must therefore also be capable of treating core positions as the rationale or justification for surrounding them with the appropriate peripheral enhancements and protections.

These demands will appear more mysterious in the abstract than in the concrete. Recall the two main varieties of rights distinguished in the previous chapter. The core of a liberty-right consists of a (full) liberty, while its periphery consists of further elements (especially claims, but also powers and immunities) whose function is to enhance and protect the exercise of the core liberty. The core of a claim-right, on the other hand, consists of a claim, while its periphery consists of further elements (especially powers, but also liberties and immunities) whose function is to enhance and protect the core claim. Examples of both varieties are abundant in legal systems. Liberty-rights, for instance, include the familiar civil liberties such as freedom of speech and association, as well as property rights, while claim-rights include the equally familiar welfare rights such as health care and education, as well as contractual rights.

Examples of both varieties are equally abundant in non-legal systems. Consider, for example, the case of a university. Students at the University of Toronto have the right to form clubs or other associations for a wide variety of political, educational, and recreational purposes. This right is a liberty-right, since its core consists of the (full) liberty both to form an association and not to do so; neither option is prohibited by the rules of the university. This core is in turn protected by rules which prohibit disrupting or otherwise impeding the activities of the association, thus imposing duties on other members of the university community, and by rules which entitle the association both to a share of funds appropriated for such purposes and to the use of university premises, thus conferring claims against the university on the members of the association. Furthermore, were we to dig further we would unearth powers (to waive the aforementioned claims), immunities (against the alienation of these various advantages by others), and doubtless higher-order liberties and claims as well. The right of University of Toronto students to form associations thus consists of a core liberty which defines the content of the right and a periphery which enhances and protects the exercise of that liberty. The rule system of the university is rich enough to treat the core liberty as the rationale for the addition of the further elements.

Students at the University of Toronto also have the right to be graded fairly and without undue delay. This right is a claim-right, since its core consists of a claim against their teachers, whose conduct is constrained by the university's rules governing grading practices. This core is in turn protected by rules giving students recourse to procedures of complaint and appeal when they feel that they have been treated unfairly, thus conferring on them both powers and the liberties to exercise those powers or not as they choose. Once again, were we to inquire further we would doubtless unearth immunities (against the alienation by others of these various advantages), as well as further claims, powers, and liberties. Again the point is that the right of University of Toronto students to fair grading is not just a random assortment of Hohfeldian advantages. Instead, it consists of a core claim which defines the content of the right and a periphery which enhances and protects that claim. Once again the rule system of the university is capable not only of conferring and imposing these various positions, but also of treating some as the point or justification for the creation of others.

These mundane examples from the rule system of a university could easily be replicated in other institutional rule systems. Whether a given legal system confers a given right on those to whom it applies is a question of the content (and scope) of its rules, and that is, broadly speaking, an empirical question. Likewise, whether a given institutional rule system confers a given right on those to whom it applies is a matter of the content (and scope) of *its* rules, and that too is an empirical question. A legal right is a right recognized within some existing legal system. An institutional right is a right recognized within some existing (legal or non-legal) institutional rule system. Just as legal rights are conventional, so are institutional rights conventional. A legal system is merely the most highly developed variety of institutional rule system, and legal rights are merely the most highly developed variety of institutional rights.

3.3 NON-INSTITUTIONAL RIGHTS

The first step toward broadening the existence conditions for legal rights is now complete. The second and final step will carry us beyond institutional rights to the general category of conventional rights. The main task here is to show how rule systems which are non-institutional, but still conventional, can also confer rights. The absence in such systems of specialized legislative and adjudicative institutions deprives us of the convenient structural parallels which hold between legal and non-legal institutional rule systems. In particular, where a rule system lacks a specialized body authorized to interpret and apply its rules we will need some other means of determining the content and scope of those rules. We will also need some substitute for the notion of systemic validity, since a rule system without adjudicative institutions cannot contain what we have thus far understood as a rule of recognition. What sense, indeed, will it make to speak of a *system* of rules in the absence of the unity imposed by a rule of recognition addressed to adjudicative officials? And even if we can make sense of speaking here of a rule system, how could we ever establish its efficacy in the absence of those officials whose acceptance of the rules plays such a prominent role in the existence conditions for institutional systems?

Perhaps the appropriate conclusion is that there can be no non-institutional rule systems, and thus no non-institutional rights. Perhaps the notions of a rule and a right are applicable only where we can locate a formal structure consisting of specialized legislative and adjudicative bodies. Such despair, however, seems premature.

For there appear to be instances of social groups which, while lacking this institutional structure, manage none the less to possess rules. Since we cannot call these groups institutions, let us instead call them associations. Some of them are the less formal counterparts of institutions: teams, clubs, interest groups, political movements, and the like, which have simply never developed specialized bodies for creating or applying their rules. Others are groups for which the absence of an institutional structure is the norm: families, communes, street gangs, peer groups, car pools, committees, and so on. It seems rather imperious to decree that associations of these sorts cannot possess rules applying to their members, especially since many of them plainly consider that they do. The more prudent course seems to be that of trying to construct existence conditions for these less structured social groups. While an institution necessarily possesses rules, it seems that an association may or may not do so, or (which is not the same thing) may or may not purport to do so. An account of the existence conditions for non-institutional (but conventional) rights will therefore need to show (1) what it is for an association to exist, (2) what it is for an association to possess rules, and (3) what it is for such rules to confer rights on those to whom they apply.

Every association is a set of individuals, but not every set of individuals is an association. Two conditions seem to be individually necessary and jointly sufficient in order for a set of individuals to form an association. The first is that the various members of the set share some common goal. There are no restrictions on the content of this goal except that it be intelligible as a practical objective, thus as the rationale of some activity or course of action. It may therefore consist of anything from the building of a utopia to the exchange of second-hand auto parts. All that is necessary is that we be able to identify some common objective which is shared by the various members of the set.[28] It is obvious that satisfaction of this condition, while it seems necessary for the cohesion characteristic of associations, is not sufficient. It may be true that all the chess players in Hartlepool have the same goal, namely finding opponents for chess matches, but this fact alone does not form the chess players of Hartlepool into an association.

---

[28] An objective may be common in this sense while also agent-relative. It is enough that each member of the set wish to engage in the same (generic) activity, even if no one cares, for its own sake, whether the others do so. Common objectives may of course also be agent-neutral.

We must therefore add a second condition. We can begin by stipulating that the various individuals who share a common goal be aware of this fact, thus that each know that (at least some of) the others are also members of the set. Now the existence of the set of goal-sharers is itself common knowledge among its members. Then, since this common knowledge is still insufficient to form the individuals into an association, we must further require that each member of the set intend to pursue the common goal in concert with (at least some of) the others. It is this common intent which converts a common goal into the collective pursuit of that goal. An association thus requires a level of interdependence and co-ordination on the part of its members, who must regard themselves as united in a common enterprise. This pattern of interlocking and mutually reinforcing beliefs and intentions is of course merely another case of the interdependence and co-ordination which is necessary for the existence of a convention.

A common goal plus the common pursuit of that goal confers an identity upon an association. This identity in turn enables us to explain what it is for an association to come to exist, to endure, to change, and to pass out of existence. An unassociated set of individuals is nothing more than its members, and thus changes its identity with every change in its membership. An association, by contrast, has a life of its own; its *raison d'être* enables it to endure through fluctuations of membership, as long as new recruits continue to acknowledge its goals and to share a commitment to their joint pursuit.[29]

If we think of primary rules as imposing constraints on the activities of those to whom they apply then an association need not possess such rules. If the group is of one mind concerning its goals and also concerning the means of their joint pursuit then each member may play his/her part willingly without the necessity of constraints. Indeed, some associations, such as communes, may exist for the express purpose of sharing a rule-free form of social existence. (But is it a rule in such communes that there are to be no (other) rules?) Additional conditions must therefore be satisfied if an association is to possess rules. In formulating these conditions it will be convenient to focus on the special case of primary rules

---

[29] The full story concerning the identity criteria for associations will doubtless be very complicated, but we can get by with the simple story.

which regulate the activities of members of the association. When should we say that an association possesses such rules?

Some of the needed conditions can be readily adapted from our notions of conformity, compliance, and acceptance.[30] Suppose that we wish to know whether the members of some association have a rule requiring some particular form of conduct. The weakest evidence in favour of the existence of such a rule would be that these members by and large engage in that conduct on the appropriate occasions. From this uniformity of conduct alone, however, we may not infer the existence of a rule; the members of the association may each eat three meals a day without there being any rule of the association requiring them to do so. Stronger evidence would therefore be furnished by the members' reasons for engaging in the conduct in question, thus by a particular explanation for the uniformity. As a minimum we would expect members to connect the conduct to the common pursuit of the group's common goals, thus to engage in it *qua* member of the group. But this too will be insufficient, since each member may elect the necessary conduct quite willingly, thus obviating the need for the sort of constraint imposed by a rule. Thus it must also be generally true that members engage in the conduct in question at least in part because of social pressure on the part of other members to do so. This pressure will standardly take the form of approving and positively reinforcing the conduct in question, and of disapproving and negatively reinforcing its omission. The positive reinforcement may take the form of any incentive ranging from praise to more tangible rewards, while the negative reinforcement may likewise range from criticism through harsher sanctions. Thus we now have the condition that the uniformity of conduct exists at least in part because of a general pattern of social pressure which tends to channel behaviour in that direction. From mere conformity we have moved to compliance.

In an institutional rule system private members need do no more than comply with its rules, since we can look to the institution's officials for acceptance of those rules. These officials will also serve as the sources of at least some of the sanctions which reinforce conformity with the rules. Since this resource is, by definition, lacking in an association, social pressure must in this case be

---

[30] The account which follows depends heavily on Hart 1961, 54 ff.

applied generally by members of the group. Thus it must be generally true that members treat engaging in the conduct in question as a standard of desirable or acceptable behaviour. The usual sign of this attitude toward the conduct will be the use of an evaluative or normative vocabulary, so that conforming behaviour is characterized as right, or proper, or normal, while nonconforming behaviour is described as wrong, improper, or abnormal. The use of this vocabulary expresses approval of conformity and disapproval of deviance, thus furnishing some of the social pressure in favour of the former and against the latter. Attitudes of approval and disapproval, acceptance and rejection, sympathy and hostility, gratitude and resentment, and the like, will in turn often be accompanied and further reinforced by other forms of social pressure. Those who deviate from the norm may therefore be ostracized, expelled, retaliated against, or even persecuted. The disapproval of deviant behaviour which is expressed by an evaluative vocabulary is inseparable from the use of informal sanctions to encourage conformity and discourage nonconformity. The widespread resort within an association to reinforcement mechanisms such as praise and criticism is the surest sign of the existence within that association of rules regulating conduct. But that is to say that such rules exist within the association in virtue of the fact that they are generally accepted, and that an association's rule system consists of all of its generally accepted rules. The unity of such a system, which entitles us to call it a *system*, will thus be a function of two related factors: the function of the rules in advancing the common goals of the group and the common, or at least general, acceptance of the rules by the members of the group.

Let us suppose that something like the foregoing account is adequate to explain what it is for an association to have a set of regulative rules. The fact remains that not just any set of such rules is capable of conferring rights on (some or all) members of the association. If we once again employ the conception of rights as protected choices then in order to confer rights a rule system must be capable of conferring liberties and claims on the one hand, and powers and immunities on the other. There seems little doubt that institutional rule systems are capable of speaking the modal languages necessary for recognizing rights; it remains to be shown that non-institutional systems can also do so. Here it will be convenient to adopt a simplification suggested by the Hohfeldian analysis of the previous chapter. In distinguishing first-order from second-order rules, I suggested that the basic pragmatic function of the former is to impose duties while that of the latter is to confer powers. In each case it is the exercise of this function which alters the normative *status quo ante*. Thus if we can show that the rules of an association are capable both of imposing duties and of conferring powers, we will have done enough to show that they are capable of speaking the language of rights.

Thus far I have assumed, for convenience, that a duty is simply whatever is required by a rule whose function is to regulate conduct. This assumption was safe enough as long as we confined ourselves to legal rules, but its deficiencies quickly become apparent when we apply it to other varieties of rule systems, and especially to non-institutional rule systems. Some rules of conduct within such systems do not impose duties (or obligations) but instead set out what agents ought to do, or what it is desirable for them to do, or what it is meritorious or praiseworthy for them to do, or whatever. If we still speak loosely of such rules as formulating requirements they will be the requirements of virtue or merit rather than those of duty. We need, therefore, a further account of when a rule of conduct which is in force within an association imposes duties on those to whom it applies.

The most important signs that a rule imposes a duty appear to be the following.[31] First, the social pressure backing the rule is especially strong or insistent. Thus conformity with the rule is generally thought to be important within its particular domain. This relativity of the weight of a duty to a domain or social practice must not be overlooked, for otherwise it would be impossible to explain how duties, like rights, can be trivial. Consider, for example, the duties imposed on hosts by the rules of etiquette. Most of us would regard these social requirements as slight when they compete with other sorts of consideration. But they do not always so compete, and the requirements can be taken sufficiently seriously within their own realm to constitute duties. The second sign concerns the polarity, as it were, of the reinforcing pressure. Rules which formulate ideals or standards of excellence are typically backed by positive inducements such as public praise, monetary

---

[31] I here borrow from Hart 1961, 84 ff.

rewards, medals, and statues on prominent sites. By contrast, rules which impose duties are typically reinforced by negative sanctions such as shame, dishonour, humiliation, or worse. In the former case failure to meet the standard is taken to be the norm and the exceptional success is therefore celebrated, while in the latter case success in meeting the standard is taken as a matter of course while failure is condemned.

When the rule system of an association contains rules whose general acceptance and application display these characteristic features then it is capable of imposing duties on its members. If they are to possess rights, however, it is also necessary that the rule system be capable of conferring powers. An institutional rule system necessarily has this capacity, since it necessarily possesses legislative and adjudicative officials. But a non-institutional rule system, even one capable of imposing duties, may be incapable of conferring powers. A power is the ability to control or manipulate normative relations. A rule system which imposes duties on its members will lack rules conferring powers if the members of the system, both individually and collectively are unable to alter those duties. This might be true, for instance, in an association entirely ruled by the dead hand of tradition, in which it would be an affront to suggest modifying what one's ancestors have dictated. Conversely, a rule system will have rules conferring powers if its members, either individually or collectively, are able to exercise control over the rules which regulate their conduct. This power would be collective if the members acting in concert could change the rules of the system. It would be individual if they were able, by some recognized means, to assume duties on their own part or to release others from their duties. In either case, whether the power exists or not would be determined by reference to the same general practice of acceptance of the requisite rules. In general, the members of the group, collectively or individually, have certain powers if the members of the group generally accept that they do, thus if they generally acknowledge and recognize the exercise of those powers. What will count as exercising a particular power will thus be determined by what is so acknowledged by the membership as a whole. In this way the members of an association, through the appropriate patterns of response and reaction, can make it true that their rule system confers powers. Just as what is permissible under the rules of the system is ultimately determined by the attitudes and practices of the members of the association, so what is possible under the rules is determined in the same way.

More is still needed in order to confirm that the rules of an association might confer rights; we have not yet established, for instance, that they are capable of imposing relational duties, and thus of conferring claims, nor that they are capable of uniting Hohfeldian positions into a coherent bundle. But it is obvious by now that the basic story will remain the same for all of these additional ingredients, and it would be tedious to extend the telling of it. Instead, I shall offer two examples of non-institutional (but conventional) rights, once again assuming the choice model and the distinction between liberty-rights and claim-rights. Consider this time the case of a family. While authoritarian family structures can locate the power to make and apply their rules solely in the parents (or, more commonly, solely in the male parent), thus converting the family into a small-scale institution, more communal arrangements are also possible. Assume that within a particular family decision-making is more widely dispersed, with both parents and children (over some minimum age, perhaps) having a say in determining the rules which all are prepared to follow. The children of that family might have, by general agreement, the right to participate in the making of family decisions. If so, then this right is a liberty-right, since its core consists of the (full) liberty either to participate or not, as the child chooses. (If the children lack the liberty not to participate then their 'right' is better described as a duty.) Does it have a periphery? Well, we would normally expect the liberty to be protected by imposing on others a duty to respect the chid's opportunity to speak during family meetings and to take the child's opinion seriously when decisions were reached. It might be protected further by enabling the child to lodge complaints in the event that this duty is breached, or to waive the protection afforded by the duty. It might be protected yet further by preventing others from exercising any of these powers on the child's behalf. If the protection of the child's core liberty extends even this far—if, that is, these further layers of protection are recognized and respected by the others—then we already have an example of a complex right with a defining core and a protective periphery.

The children of the same family might also enjoy the right to receive on a regular basis a stipulated sum of spending money. If so, then this right is a claim-right, since its core consists of the claim to this sum, presumably held against the parents. Again we should expect the core claim to attract a periphery of connected advantages which enhance and protect it. Thus we should expect that the child will have considerable liberty in spending the money, plus the power to waive payment of it, plus the immunity against payment being waived by others, plus . . . Once again, there is no impediment in principle to the child's enjoying a right with all of the complexity and coherence typical of claim-rights. All that is necessary is for the rules of the family to make it so, and for these rules in turn to be generally accepted by the family members. Since what is true of families will be equally true of other associations, there is no reason to believe that associations are in general incapable of conferring rights upon their members. Ultimately, as in the case of institutional rights, whether a particular association does so will be determined by the semantics of its rules and by conventional social practices of complying with and (especially) accepting those rules. The rights conferred by associations are therefore also conventional rights.

The generalization of our existence conditions for legal rights is now complete. Legal systems are a special case of institutional rule systems, and institutional rule systems are a special case of conventional rule systems. Legal rights are therefore one species of institutional rights, and institutional rights are in their turn one genus of conventional rights. It is this notion of a conventional right which has been our quarry in this chapter. With a reasonably full account of the existence conditions for conventional rights in hand, our next task is to develop a similar account for moral rights.

Before turning to that task, however, we should take note of the fact that, in one sense of the notion of a moral right, the existence conditions for conventional rights are also existence conditions for moral rights. The conventional morality of a society may itself constitute a non-institutional rule system, and thus may be capable of conferring conventional moral rights on those to whom it applies. For expository convenience I will focus here on the case of the homogeneous society with a settled, dominant moral code. This code may well take the form of, or at least include, both rules which regulate conduct and rules which enable those forms of regulation to be controlled and manipulated. Thus the rules of a conventional morality may, and typically do, confer rights. Furthermore, these rights may display all of the articulation demanded by the conception of rights as protected choices. Liberty-rights which are respected in many (though not all) social moralities include a domain of privacy into which others may be invited but must not intrude, while the clearest examples of claim-rights are those which are created by the conventional rules concerning promises and agreements.

In the case of a conventional morality it is probably misleading to think of it as the rule system of a particular association. Although it is the rule system of a particular society or culture, a social morality shares the comprehensiveness of a legal system and is thus not characterized by any specific goal or purpose. If we are to understand the features peculiar to a conventional morality it is probably easiest to contrast it with its legal counterpart.[32] The two kinds of rule system share some important structural features: both claim comprehensive, supreme, and compulsory authority. They are also likely to overlap to a considerable extent in content: both will regulate modes of conduct deemed to be important to the successful functioning of a society. The principal differences between them stem from the fact that a legal system is institutional while a social morality is not. Because a legal system contains specialized legislative and adjudicative bodies its rules can be authoritatively revised and interpreted in a way which will generally be unavailable to a positive morality. As a result legal rules are likely to be both more complex and more determinate than conventional moral rules. We must not overstate this contrast. A particular social morality may be quite definite in its demands concerning, say, sexual behaviour or conjugal relations. But where the rules are not definite, or where novel social conditions give rise to hard cases, then no recourse may be available to a body of officials whose judgement will be authoritative.

In lacking rules conferring public (legislative and adjudicative) powers, a social morality of course also lacks a rule of recognition. There is therefore no test of systemic validity for its rules. In the case of a legal system it is possible for a particular rule to be valid even though it is widely rejected, ignored, or flouted by those to whom it applies. In a shared social morality the rules 'in force' can be only those which continue to command general allegiance. The validity of a particular rule is thus a direct, though also complex,

---

[32] See Hart 1961, 168 ff.; Wellman 1985, 111 ff.

function of its efficacy. This feature is, of course, shared by all non-institutional rule systems.

The non-institutional nature of a social morality is also reflected in its use of sanctions. It is common to both legal systems and positive moralities that their rules are typically backed by the threat of sanctions. In a legal system the carrying out of this threat is the function of the courts and their ancillary law-enforcement agencies. The procedures for determining guilt and imposing penalties are formal, the schedule of penalties for a given offence is determinate, and the penalties themselves may be severe. A social morality, by contrast, contains no special apparatus for enforcing its rules. As a result, enforcement tends to be local, uncertain, and informal— though wider and more organized compaigns of both glorification and vilification are certainly possible. It is also unusual, though far from unknown, for societies to reinforce their conventional morality by resorting to the harsher penalties of confinement, physical injury or death. The more common (negative) sanctions are expressions of disapproval or contempt, ostracism, humiliation, and the like. Further, because a social morality only exists if its rules command the hearts and minds of members of society, enforcement of the rules tends to rely heavily on their internalization, and thus on appeals to the 'internal sanctions' of guilt and shame.

Finally, a legal system and a positive morality are also likely to display some differences of content. In a legal system rules regulate behaviour basically by imposing duties; there is no place in such a system for standards of noble or meritorious conduct. A social morality, by contrast, will typically include such standards. It is also typical of a social morality that it will concern itself solely with those modes of conduct regarded as important for the smooth functioning of society. A legal system, by contrast, will regulate the trivial as well as the weighty, and will thus have something to say about many matters on which positive morality is silent.

In most societies one legal system and one or more social moralities exist side by side. Since they concern themselves with many of the same modes of conduct, there will inevitably be complex causal relations between them. There will also be instances in which the two systems stand in open conflict, the legal treatment of some mode of conduct being condemned by conventional morality. Each rule system may confer rights on those to whom it applies, and the same conception of a right and the same set of existence conditions for rights will enable us to determine, in principle at least, the rights conferred by each. There is thus no theoretical impediment standing in the way of identifying the rights enshrined in the conventional morality of a given society, or of distinguishing between those rights and that society's legal rights. In this sense moral rights are not particularly mysterious.

But conventional moral rights are plainly not the moral rights whose existence conditions we are seeking. Moral rights, in the sense which is of interest to us, necessarily have moral force. Thus, wherever a moral right is applicable its existence must count as a moral consideration in favour of some courses of action and against others. This consideration need not, of course, be decisive or conclusive; moral rights often compete with moral reasons of other sorts. But where a moral right is genuine, and where our conduct falls within its domain, its existence must be accorded some deliberative weight. Conventional rights, of whatever variety, have their own kind of normative force. Thus legal rights have legal force; that is, they count as reasons for action under the rules of a legal system. The same will be true of the rights conferred by other conventional rule systems, both institutional and non-institutional; all have the status of reasons within their own system. Conventional moral rights therefore also have normative force, since they function as reasons within the framework of a shared social morality. But the source of the normative force of rights in the case of all conventional rule systems lies entirely in the background social practices of compliance and acceptance which sustain those systems. Conventional rights count as reasons within their respective domains because they are commonly counted as such; remove their foundation of widespread allegiance and endorsement and their force will evaporate.

In all of these respects the normative force of moral rights is different. That force does not depend on recognition or acknowledge-ment of the rights within any conventional rule system or by the members of any institution or association. Or so we are assuming. It is this independence of conventional recognition which enables moral rights to set standards of justice for the design of conventional rule systems. If a particular system meets these standards then the rights which it confers may have moral force. However, because many systems conspicuously fail to meet these standards many conventional rights lack such force. Whether a conventional right has moral force is therefore a contingent matter to be determined, at least in part, by reference to some rights which are not themselves conventional. It follows that if there are any genuine moral rights their existence conditions cannot be those of conventional rights of any kind, including conventional moral rights.

4

Natural Rights

IF moral rights are not conventional rights then what are they? One influential and appealing answer to this question has been offered by the moral tradition which affirms the existence of natural rights—or natural justice, or natural law. In effect, the natural rights tradition attempts one possible extrapolation from the existence conditions of conventional rights to those of moral rights. Conventional rights are the products of conventional rule systems. Suppose that we extract from this result the idea that all rights are the products of rule systems of some sort or other. Since conventional rules can confer only conventional rights, the rules needed to confer moral rights cannot be conventional. But if they are not conventional then it seems reasonable to think that they must be natural. Thus are we easily led to the supposition that moral rights are the products of a system of natural moral rules (or laws), thus to the supposition that moral rights are natural rights.

The intuitive plausibility of this supposition lies mainly in the elegant analogy which it constructs between conventional rights (of whatever variety) and moral rights. Just as there are conventional rule systems which apply to various groups of us, so there is a system of natural rules which applies to all of us. Just as conventional rule systems are capable of speaking the modal languages necessary for imposing duties and conferring powers, so is this natural system. Thus, just as conventional rule systems are capable of conferring conventional rights, this natural system is capable of conferring natural rights. However, unlike conventional rights these natural rights will have moral force, since the rules which generate them are themselves moral rules. The moral force of moral rights can thus be captured and explained by the moral force of natural moral rules.

One attractive feature of this structural analogy is that it enables us to maintain, indeed extend, our provisional taxonomy of rights. As initially proposed, that taxonomy held that rights constitute a genus of which the several particular varieties of rights are so many species. Our analysis of conventional rights has merely complicated this scheme by requiring the introduction of a higher level of classification. Now we have come to treat conventional rights as a biological family with two genera—institutional rights and non-institutional rights—each of which subdivides in turn into various species. The natural rights tradition fills out this picture by treating moral rights as a complementary family, perhaps with interesting subdivisions of its own.

If conventional rights and moral rights are the two families of a common order then they must share a common concept of a right. But the existence conditions for the two families must also be analogous. Since the natural rights tradition accepts the assumption that all rights are the products of rules, the existence conditions for any particular variety of rights will depend on the existence conditions for the rule system which generates them. Thus if the natural rights strategy is to succeed it must be able to provide existence conditions for natural moral rules (or laws). Furthermore, these existence conditions must be capable of preserving and explaining the moral force of moral rights. In the existence conditions for conventional rights the normative force of these rights (within their particular rule system) can be entirely explained by reference to general social practices of compliance and acceptance. Since these practices are insufficient to generate moral force, the existence conditions for moral rights will need to appeal to some other factor. The natural rights tradition seeks to explain the moral force of rights by embedding them in a system of rules which, by virtue of being natural rather than conventional, itself has moral force. Thus the main task for the tradition will be to provide existence conditions for a rule system which is both natural and normative.

4.1 THE NATURE OF NATURAL RIGHTS

If we are to assess the central claims of the natural rights tradition we will need a better understanding of them. What, then, makes a right a natural right? Unfortunately, there is no unique answer to this question in the tradition. The existence of natural rights has been affirmed both in political rhetoric and in political theory. Rhetorical assertions of such rights have usually taken the form of declarations or manifestos in the service of some political cause. These documents are not remembered chiefly for the depth of their philosophical reflection. Declarations and manifestos seldom offer a grounding for their catalogues of rights, and they never offer an account of what makes a right a natural right.

What then of natural rights theories? The tradition may be usefully divided into its classical and modern periods. The heyday of the classical period was the seventeenth century which featured the great treatises by Grotius, Pufendorf, and Locke.[1] These writers, and especially Locke, are generally regarded as having laid the theoretical foundations for the revolutionary rights rhetoric which characterized the last quarter of the eighteenth century. Certainly the tradition's classical period had come to an end by the latter part of the nineteenth century, when natural rights theories were eclipsed in their British homeland by the rise of utilitarianism on the one hand and idealism on the other. Interest in natural rights then remained relatively dormant until the Second World War, after which it underwent a marked revival, at least in Anglo-American philosophy and politics.

Neither the classical nor the modern period, however, yields an agreed and considered account of the nature of natural rights. While the classical theories appear to share a tacit understanding of what it is for a right (or a duty or a law) to be natural, this understanding remains too undeveloped to serve our purposes. The modern era, by contrast, features a great many philosophical explorations of natural rights but yields no consenus. Despite all the discussion there is no agreement, for instance, on such basic issues as whether natural rights are alienable, prescriptible, forfeitable, defeasible, or self-evident. Furthermore, it is a curiosity of the modern debate that only scant attention has been given to the question of what makes such rights *natural*.[2]

Since the tradition offers no account of natural rights which we can simply take over, we must construct our own. Clearly, however, we may not have our own way with this important notion. Because it will be particularly important not to characterize natural rights in some question-begging or idiosyncratic manner, I

---

[1] For a history of natural rights theories see Tuck 1979.

[2] The most influential discussions of natural rights in the modern period have been Macdonald 1946–7; Brown 1955; and Hart 1955. But there have also been numberless others.

will require our account to satisfy two criteria. The first is that it be as faithful as possible to what is commonly understood as the natural rights tradition. Since there is no agreement within that tradition concerning some of the features of natural rights, we cannot require that every ingredient in our account be universally accepted. But we can, and should, require that each ingredient be at least widely accepted. It may be possible for us to remain agnostic on some issues which deeply divide the tradition. Where this is not possible and we are compelled to take sides, we will want to offer a good reason in favour of our preferred option.

The second criterion of adequacy is that our account should enable us to mark some important lines of division between different types of moral/political theory. Whatever natural rights may turn out to be, their affirmation appears to constitute one of the main options open to us in constructing a general normative theory of morals and politics, and especially in constructing a theory of justice. An account of the nature of natural rights should therefore illuminate the deep differences between natural rights theories and their principal rivals.

The account which follows gives prominence to the second criterion by pursuing a two-stage strategy. My primary aim will be to determine what is to count as a natural rights theory; I will then characterize natural rights in terms of their role in such a theory. The effect of this strategy, therefore, is to treat natural rights as theoretical entities; what makes them natural is the niche which they occupy in a particular moral structure. My answer to the primary question will consist of four conditions which are individually necessary and jointly sufficient. As a rough initial approximation we may say that a moral theory is a natural rights theory just in case (1) it contains some moral rights, which (2) it ties to the possession of some natural property, and which it treats as both (3) basic and (4) objective. Each of these four conditions requires explication.

The first condition is conceptual: a natural rights theory must affirm the existence of some moral rights and thus must employ some conception of a right. This will seem so obvious as to be scarcely worth stating—surely a natural rights theory must at least be a rights theory—and it is in any case entailed by the next two conditions. The point of stating it separately is to focus attention on a conceptual issue which we have thus far side-stepped. We have in play two different models of a right, centred respectively on interests and on choices. On each of these conceptions the function of a right is to protect some value on the part of the right-holder, and it is this function which unifies what would otherwise be disparate Hohfeldian atoms into a cohesive molecular structure. However, because the value to be protected differs on the two conceptions (welfare on the one hand, autonomy on the other), what the two conceptions will count as rights will correspondingly differ. Because autonomy can be treated as a particular component of individual welfare, anything which counts as a right under the choice conception will also count as such under the interest conception. But the reverse will not be true, since there are ways of protecting welfare without protecting autonomy.

Thus far we have managed to remain agnostic concerning the relative merits of these two conceptions. As I noted at the end of chapter two, there is probably no way of settling the issue between them if we confine ourselves to a standard of descriptive adequacy, since neither account will fit all of our pre-theoretical intuitions about rights. On the one hand the interest conception threatens to acknowledge some right-holders who are not generally counted as such (i.e., third-party beneficiaries), while on the other hand the choice conception threatens to disqualify some right-holders who are generally counted as such (e.g., children). Where legal systems are concerned the outcome is pretty well a stand-off, and it remains so when we expand our range so as to include other sorts of conventional rule system. In the absence of any compelling reason for preferring one conception to the other we have therefore retained both, merely using the choice conception for illustrative convenience when nothing in the argument depended on taking sides.

However, we now have a further standard of theoretical adequacy in terms of which we can compare the merits of the two conceptions. We are seeking a characterization of natural rights theories which, among other things, illuminates the important lines of division between them and our other theoretical options. Whereas the requirement that such theories contain some conception of a right is trivial, the selection of a particular conception may not be trivial. Since the two alternatives are not extensionally equivalent they are bound to yield different maps of the theoretical terrain. If one of these maps identifies more significant theoretical boundaries than the other, and if it seems advisable to use the concept of a right to mark these boundaries, then we will have good reason for preferring the conception which yields that map. Now it seems to me that the map generated by the choice conception does identify more important theoretical boundaries and that we would do well to treat these as the boundaries between rights theories and their rivals. If this is so then we have a good theoretical reason for favouring the choice conception.

The basic difference between the two conceptions lies in the normative function which they assign to rights. On the interest conception that function is the protection of some aspect or other of the right-holder's welfare. While this function might be most efficiently served in some cases by assuring the right-holder's liberty or normative control, in other cases it might be most efficiently served by restricting that liberty or withholding that control. Thus, for example, certain basic interests of individuals might best be served by conferring on them claims to the protective services of others which they have no power to alienate, whether temporarily (by waiving them) or permanently (by relinquishing them).[3] Since the interest conception will count such claims as rights, there are no internal connections on this model between rights and such values as autonomy, self-determination, and freedom. These connections are of course affirmed by the choice conception, since on that conception rights exist precisely in order to protect and promote these values. Thus on the choice conception a claim which cannot be alienated in any way, thus one which is beyond its holder's normative control, cannot count as a right. Now the difference between a normative structure in which autonomy and its cognates are the centre-piece and a structure in which they are not is an important difference. Structures which highlight autonomy will treat individuals as active managers of their own lives even when doing so will work to their overall detriment. On the other hand, structures which highlight welfare will treat individuals as managers when that is likely to be in their interest and will otherwise treat them as the passive beneficiaries of the services of others.

It is not overstating the case to say that normative structures which differ in this fundamental way presuppose divergent conceptions both of the person and of the moral life. Because this

[3] For a valuable discussion of inalienability see Feinberg 1980, ch. 11.

difference is evidently a deep one, it is worth marking in some emphatic way. Since it is possible to build a moral theory around either the value of individual welfare or the value of individual autonomy, it is also worth marking the boundary between these theoretical options in some emphatic way. The language of rights lends itself readily to this purpose. On the one hand its prominence in moral/political debate means that any boundary drawn between rights theories and their rivals will be noticed. On the other hand we have available a conception of a right which enables us to draw exactly this boundary. If we adopt the model of rights as protected choices then we can assign a distinctive normative function both to rights and to those moral theories which, in one way or another, take rights seriously. We can say that to regard individuals as having certain moral rights is to regard them as being autonomous within the domains specified by the contents of the rights.[4] In the case of a liberty-right the core content is given by a (full) liberty which assigns the agent freedom of choice. In the case of a claim-right the core content is given by a claim which the agent has the power to alienate or otherwise manipulate. On the choice conception every right must contain either liberties or powers (or both), and thus must confer either freedom or control (or both). The function of the peripheral elements of the right is then to enhance and protect this freedom/control, and thus the agent's autonomy. Rights on the choice conception thus enable us to distinguish effectively between two grounds for imposing constraints on others: the protection of individual autonomy and the protection of individual welfare.

The proposal to adopt the choice conception is therefore based on two contentions: that the concept of a right is sufficiently important to be assigned a distinctive normative function, and that autonomy is sufficiently important to be safeguarded by a distinctive normative concept. Were we to adopt the interest conception instead we would be unable to use this important concept to safeguard this important value. We would, of course, still be able to distinguish between protecting individual welfare and protecting individual autonomy. But the distinction would be drawn within the domain of rights rather than between that domain and its neighbours. Furthermore, there would seem to be

[4] My notion of autonomy within a domain is essentially similar to the notion of dominion in Wellman 1985, 95 ff.

no effective way of highlighting the distinction. Since the content of a right is given by its core, and since on the interest conception the right-holder may or may not have normative control over the core of a claim-right, even the content of a right does not always discriminate between those cases in which agent autonomy is assured and those cases in which it is not. The upshot is that if rights are construed as broadly as the interest conception dictates, the welfare/autonomy distinction will largely drop from sight.

However, the interest conception does in its own way assign a unique normative function to rights. Since there can be many grounds for imposing constraints on others, specifying that the ground in a particular case is the protection of some individual interest singles out an important kind of justification. Since the interest conception construes rights as having the function of protecting interests, it can mark this range of cases by saying that here the constraint is grounded in the beneficiary's right.[5] Thus the interest conception does not entail that rights are functionally redundant, nor that their domain is identical to that of duties. Instead it offers a competing view of the distinctive function of rights. One way to arbitrate between the two conceptions would be to argue that the theoretical boundary yielded by the one is more important than that yielded by the other. But it would be a large task to show that the distinction between the protection of individual autonomy and the protection of individual welfare is more important than the distinction between the protection of individual welfare and other grounds for imposing constraints.

There is in any case another way of deciding the issue. Just as we should avoid rendering an important moral concept functionally redundant, we should also avoid unnecessary duplication of function between different concepts. Thus if we are deciding whether to mark a particular boundary by means of the language of rights we should ask ourselves whether we have the resources to mark it in some other way. I argued above that if we do not use the concept of a right to mark the autonomy/welfare distinction we will be hard pressed to mark it in any other striking way. However, we do have available an alternative way of marking the distinction between the protection of individual welfare and other grounds for imposing constraints. We cannot simply use the language of duties

---

[5] As does the account in Raz 1984(*a*); cf. Dworkin 1977, 171–2.

for this purpose, since duties may have many different grounds. But we can use the language of relational duties. A relational duty is a duty owed to some specified party who holds the correlative claim. Earlier we distinguished two different analyses of the relationality, or directionality, of relational duties. On the benefit analysis a relational duty is owed to the party who is its intended beneficiary, while on the control analysis a relational duty is owed to the party who has the power to manipulate it (by annulling it, postponing performance of it, seeking remedies for non-performance, and so on). Like the competing models of rights, these competing accounts of relational duties identify different rationales for imposing duties: the protection of individual welfare and the protection of individual autonomy. Therefore, if we choose not to use the concept of a right to highlight those constraints whose function is to protect individual welfare, we can do so instead by using the concept of a relational duty.

To see the ways in which our conceptual resources might be utilized consider the following three grounds for the imposition of constraints: (1) the protection of individual autonomy, (2) the protection of individual welfare, but not autonomy, and (3) the general welfare. Plausible examples of duties falling into these three categories might include duties to respect privacy, duties not to inflict injury, and duties to contribute to public goods. On the interest conception of rights the duties in both (1) and (2) have corresponding rights while the duties in (3) do not. The problem here is that we have no obvious resources for marking the distinction between (1) and (2). On the choice conception, by contrast, the duties in (1) have corresponding rights while the duties in (2) and (3) do not. However, here we can also say that the duties in (2) are relational, thus owed to the individuals which they protect, while the duties in (3) are not. Thus we can use the relational/non-relational distinction to do the work which the interest conception assigns to rights. In those cases in which a duty has the function of protecting individual autonomy, by virtue of belonging to an appropriate bundle of Hohfeldian positions, we will speak of the protected individual as the holder of a right. In those cases in which a duty has the function of protecting individual welfare, but not autonomy, we will speak of the duty being owed to the protected individual and of this individual as the holder of a claim. But we will not speak here of rights.

It may seem somewhat odd to adopt both the benefit account of relational duties and the choice conception of rights. It might be thought that the benefit account and the interest conception on the one hand, and the control account and the choice conception on the other, make more natural pairings. It is true that the route I am choosing has one slightly awkward implication, namely that I might have a duty to you while you have no right against me. This will be true in any case in which you lack the appropriate forms of normative control over the duty (thus, for example, if you are a third-party beneficiary of some agreement on my part). This result, however, is not very awkward, since we can always say instead that you have a claim against me. In any case, adopting the choice conception of rights does not require adopting also the control account of relational duties. On the choice conception every right will include some claims plus some normative control over those claims. But that control need not be entailed by the very concept of a claim, since it can be added by other elements in the periphery of the right. There is thus no inconsistency in combining the benefit account of relational duties (and thus of claims) with the choice conception of rights. Indeed, adopting the benefit account has the mild advantage of maintaining the logical separation between the first- and second-order Hohfeldian elements. But its main advantage lies in allowing us to mark one important boundary between different kinds of duties without utilizing the concept of a right.

The first condition therefore specifies that a theory contains rights only if it accepts the model of protected choices. It goes no way, however, toward explaining what it is for such rights to be natural (the model of protected choices seems, on the face of it, neither more nor less natural than that of protected interests). The second condition will take one step toward such an explanation. A theory of rights will contain not just any rights but some particular set of rights. If these rights are to be capable of performing their theoretical function then they will need to possess a reasonably determinate content, scope, and strength. Setting the first and last aside for the moment, let us focus on one aspect of the scope of these rights, namely their subjects. A subject of a right is a holder or bearer of that right, thus a member of the class of beings on whom the theory confers the right. Although it might be quite reasonable to assign different moral rights to different subjects, it will be convenient to assume, for the sake of the present exposition, that all of the rights contained in a particular theory are held by the same class of beings. In that case we face the obvious problem of determining the membership of this class. It seems reasonable to demand that any particular distribution of moral rights be justified in some way. One way to provide such a justification is to defend what we may call a *criterion* for the rights, namely some property (or set of properties) whose possession is both necessary and sufficient for possession of the rights. It is characteristic of natural rights theories to employ this line of defence.

Their adoption of this strategy does not yet provide any sense in which (at least some of) the rights affirmed by such theories might be natural. But it does suggest an obvious requirement: the criterion for a natural right must itself be a natural property.[6] A natural rights theory therefore must assign (at least some of) its rights to a class of subjects determined by their common and exclusive possession of this natural property. We need not speculate here about which property might be selected for this role. In the natural rights tradition moral rights have standardly been assumed to be human rights, and thus membership in our species has been the most popular criterion for being a right-holder. But there is nothing in the nature of a natural rights theory which requires it to settle on this natural property rather than some other.

We do, however, need to think a little about the following question: what, in this context, makes a property a natural property? Two requirements are obvious. First, the property must be empirical and thus whether or not an individual possesses it must be ascertainable by ordinary empirical means. Thus supernatural properties such as having a soul or being one of God's elect are excluded. Secondly, the property must not logically require the existence of any particular institution or social practice. Thus conventional properties such as citizenship, social status, and wealth are excluded. A natural property must therefore be both empirical and non-conventional; obvious candidates, besides membership in the species, include race and gender.

To these two requirements it might be wise to add a third. Some properties while themselves logically independent of the existence of social institutions or practices are none the less causally dependent on them. Health and longevity are obvious examples. It

---

[6] This requirement can be found in Macdonald 1946–7, 227 ff., and in Brown 1955, 196–8.

would be contrary to the spirit of a natural rights theory to tie possession of its rights to any property which can be too strongly influenced by conventional practices. This third requirement is much less clear-cut than the first two, since there are few personal characteristics which are utterly immune to social influence. But some characteristics are less immune than others, and they will not serve as criteria for the possession of natural rights.

The intuitive idea behind the requirement that the criterion for (at least some) moral rights must be natural is that such rights must not be distributed on morally irrelevant or arbitrary grounds, and that only natural properties will pass this test. However this may be, it is clear that not all natural properties will pass it. A theory which assigned moral rights to individuals on the basis of race or gender would not be a very plausible theory, but it would satisfy both of our conditions so far for being a natural rights theory. But this is as it should be; the category of natural rights theories must be constructed so as to make room for both good and bad specimens of the kind.

Thus far, then, a natural rights theory is any moral theory which contains rights whose criterion is natural, and natural rights are any such rights. These two conditions appear to be sufficient to define what counts in some influential circles as a natural right.[7] Because this sense in which a right might be natural looks entirely to the criterion for its distribution, I shall call it the purely extensional sense. This characterization of what makes a right a natural right is doubtless well-suited for some analytical purposes. But it will not suit our purposes. We are seeking a particular account of the existence conditions for moral rights, namely that which has been offered by the natural rights tradition. It is thus essential to our enterprise that we be able to pick out those theories which belong to this tradition. Natural rights theories all contain rights which are extensionally natural, but many other moral theories do as well. Indeed, as we shall see later, both contractarian and consequentialist theories can accommodate natural rights in this sense. But this fact does not make them natural rights theories; if we know anything for certain about the theories in the tradition it is that they are neither contractarian nor consequentialist. We therefore need a more robust characterization of a natural right.

[7] To wit: Hart 1955; Rawls 1971, 505 n.

The third condition is structural: a natural rights theory must treat only rights as morally basic.[8] Every rights theory will assign some set of moral rights to some set of individuals. Let us assume that any such assignment will take the form of a principle which specifies the content and scope of the rights in question. Call such a principle a rights principle. Thus, for example, the principle that all and only human beings have a right to life is a (deplorably vague) rights principle. Every rights theory will locate some set of rights principles at some level in its structure. A natural rights theory will locate (at least some of) its rights principles at the most basic level of its structure, and it will locate no other moral principles at that level. Furthermore, the rights which it embeds at this basic level will be those which it treats as natural (in the extensional sense). Let us say that one principle grounds another if it provides a justification, in whole or in part, for that other. Then a principle is basic in a moral theory if other principles in the theory are grounded by it but it is not in turn grounded by any other principle. A theory treats rights as morally basic just in case all of its basic principles are rights principles. The moral base of such a theory may contain either a single rights principle or a plurality of such principles, but it may not contain any other sort of moral principle.

The intuitive idea behind this condition is that we should think of a moral theory as a set of principles, each of which employs some moral category, which are ordered into a structure or hierarchy. Within any theory some principles will be basic while others will be derivable from them (perhaps with additional premisses). The entire theory may in turn be supported by some further, non-moral foundation. Then a theory is a natural rights theory only if rights are the moral category fundamental to the theory, thus only if the theory's base consists exclusively of rights principles. There are, therefore, various ways in which a theory can fail this condition. One is for the theory to have some non-foundational structure, thus for it to contain no basic principles at all. It might, for instance, contain several principles each of which lends some support to each of the others. If such a moral structure is possible, then a theory with that structure could not be a natural rights theory.

A more realistic possibility is that a theory will indeed contain basic moral principles but that some (or all) of them will not be

[8] See Brown 1955, 195–6; Dworkin 1977, 176–7.

rights principles. If we wished to map the various structural alternatives here we would obviously need an inventory of the types of moral principle which a theory might treat as basic. But we would also need a working distinction between moral principles and non-moral principles. After all, the third condition stipulates only that rights must be *morally* basic in a theory. This prohibits grounding rights in any other moral principles, but it does not prohibit grounding them in some non-moral principle or argument. To the extent that the boundary between moral and non-moral principles is unclear, the boundary between the moral base of a theory and its deeper non-moral foundation will also be unclear.

Since it would take us too far afield to develop a full account of what makes a principle a moral principle, I will fall back here on our largely intuitive notion of moral force. A moral principle is one kind of normative rule and moral force is one kind of normative force. A rule has normative force just in case it offers reasons for action relative to some rule system. Different kinds of normative force are therefore relative to different kinds of rule system. It follows that a rule has moral force just in case the reasons which it offers are moral reasons. In order to fill out this account we would need to characterize the differences between moral reasons and other sorts of reasons. We have done some of this work by pointing to the differences between moral reasons and the sorts of reasons offered by the rules of conventional rule systems. Thus we know that moral force is more than, or other than, legal force, institutional force, customary force, and so on. But moral reasons are not the only non-conventional reasons. The rules (or principles) of a normative theory of rationality, for example, will surely offer reasons for action, but these will not (or at any rate need not) be moral reasons. Were we to develop a full account of what makes a reason a moral reason we would doubtless need to capture the special sort of impartiality or detachment which seems to be characteristic of the moral point of view. However, even in the absence of such an account it should be clear enough for our purposes when a reason reflects that point of view. The third condition prohibits grounding rights principles in other moral principles but it does not prohibit grounding them in other normative principles, such as a theory of rationality.

Emunerating all of the moral categories around which moral principles can be formed would also be a time-consuming task, if indeed it is possible. Thus I shall here content myself with mentioning three prominent contenders. Two of them—duties and rights—have already been developed in our earlier analytic investigations. Here we need only remind ourselves of the important differences between these notions. On any plausible conception of a right, rights are more complex than duties. On the choice conception, rights are bundles of Hohfeldian positions which protect the right-holder's autonomy over some specified domain. Since duties imposed on others are ingredients in this protection, all rights will include such duties. But since it is not the function of all duties to provide such protection, not all duties will be ingredients of rights. To the extent that a moral principle imposes some duty which is not an ingredient in any right, that principle speaks the language of duty rather than that of rights. Many of the principles in traditional natural law theories fall into this category either by imposing non-relational (self-regarding) duties or by imposing relational duties whose correlative claims cannot be alienated in any way.[9] We may say that such principles impose duties, whether relational or non-relational, but not that they confer rights. One of the virtues of the choice conception of rights is precisely that it highlights this distinction between duty-imposing and right-conferring principles. It therefore also highlights the distinction between duty-based and right-based moral theories. A theory is duty-based if all of its basic principles impose duties without conferring rights, right-based if they all confer rights.

The third category is that of a goal. Because we will need to explicate this notion later in our discussion of consequentialist theories, I will say little about it now. Some state of affairs counts as a goal if its achievement or realization is valuable or worthwhile, but not obligatory or a matter of right. The fact that some project or policy would promote the state will count in its favour and the fact that it would inhibit the state will count against it. The state therefore functions as an ideal or objective by means of which the merits of alternative projects and policies can be compared. The vocabulary used to formulate goals will therefore be evaluative rather than normative or deontic—the language of the good rather than that of duties or rights. Obvious examples of goals include the eradication of infectious diseases, an end to the arms race, the

---

[9] Both sorts of duties are included in the moral theories defended in Donagan 1977 and Finnis 1980.

greatest happiness of the greatest number, and a chicken in every pot. A theory is goal-based if all of its basic principles commend the promotion of some stipulated goals.

We thus have three pure cases: a moral theory may be duty-based, right-based, or goal-based.[10] It cannot be emphasized too strongly that this enumeration is not claimed to be exhaustive; it merely picks out three possible moral structures for each of which prominent examples are ready to hand. Thus there may well be other pure cases, as well as mixed theories whose base consists of principles drawn from different categories. The third condition merely stipulates that a natural rights theory must be right-based. A theory will therefore violate this condition if (1) it has no assignable base at all, (2) it is either duty-based or goal-based, or (3) it is a mixed theory.

The fourth condition concerns what we may grandly call the ontological status claimed for a theory's basic rights. Let us say that a value is subjective if its existence is entirely dependent on some state or activity, either actual or hypothetical, on the part of some subject or set of subjects, and that it is objective if its existence is entirely independent of all such subjective states and activities. Familiar subjectivist views would then include the claim that we create values through our activities of preferring or valuing or choosing, while familiar objectivist views would include the claim that we discover values by means of observation or reasoning or intuition. Since basic moral rights are values, they might be claimed to be either subjective or objective. Each of these claims would determine a different method whereby a theory might support its basic rights principles. As we have seen, the requirement that these principles be basic precludes grounding them in any further moral principles, but it does not preclude giving them some non-moral foundation. There appear to be two models available for such a foundation. If rights are subjective then their existence is ultimately a matter not of discovery but of creation or invention. Grounding a particular set of rights principles would therefore involve showing that these principles would issue from the appropriate set of preferences or choices. Call this the constructivist model. On the other hand, if rights are objective then grounding a particular set of

[10] This taxonomy obviously owes much to Dworkin 1977, 169 ff. However, I draw the distinction between right-based and duty-based theories in a different way, since I rely on the model of rights as protected choices.

rights principles would involve showing that these principles accurately represent moral facts which exist independently of our preferences and choices. Call this the realist model.

Some types of moral theory appear to be compatible with either of these methodological models; consequentialist theories, for example, have been defended in both ways. But natural rights theories are committed to the realist model and to the accompanying claim of the objectivity of rights.[11] This objectivity is the second important respect in which natural rights are natural: besides having a natural criterion their existence can be established only by means of an argument from nature rather than from any convention (actual or hypothetical).

If we pull together the four conditions for counting as a natural right we get the following picture. A natural rights theory is any moral theory which (1) employs the model of rights as protected choices, (2) assigns some set of rights to some set of individuals on the basis of some natural criterion, (3) treats these rights (and nothing else) as morally basic, and (4) claims that they are objective. A natural right in the purely extensional sense is any right which has a natural criterion. A natural right in what I shall call the metaphysical sense is any right which, in addition to being extensionally natural, is also basic and objective. The distinctive claim of the natural rights tradition is then that at least some moral rights are metaphysically natural.

Since this is the conception of a natural right which will be subjected to critical examination in the next section, we would do well to assure ourselves that it is not idiosyncratic or question-begging. How well, then, does it satisfy our two criteria of adequacy? It should come as no surprise that it enables us to locate some of the deep divisions among different types of moral theory, since it has been constructed with this aim in mind. But it is worthwhile cataloguing this theoretical pay-off. We know that natural rights theories are neither consequentialist nor contractarian. Consequentialist theories, by virtue of their goal-based structure, fail the third condition while contractarian theories, by virtue of their constructivist methodology, fail the fourth. But we also have good reason to distinguish between natural rights theories and what we might call natural duty theories.[12] Since both confer their

---

[11] See Macdonald 1946–7; Mackie 1978; Martin and Nickel 1980.
[12] I owe this perspicuous label to Peter Danielson.

favoured brand of moral protection on the basis of some natural criterion and since both also accept a realist methodology, both can lay claim to being (metaphysically) natural. But if we adopt the model of rights as protected choices then many duties will not be ingredients of rights. Any theory whose basic principles all impose such duties will be duty-based rather than right-based. The first and third conditions thus mark the boundary between theories of natural rights and theories of natural duties.

The results of applying the other criterion are more ambiguous. Three of the four conditions exclude some theories whose authors have regarded themselves as belonging to the natural rights tradition, or have been so regarded by others. The first condition excludes the theories which dominated the tradition's classical period. All of these theories were oriented toward protecting not individual autonomy but individual welfare. This orientation led them both to impose self-regarding duties, thus restricting the liberty of individuals in the private sphere, and to confer inalienable claims, thus restricting control by individuals over their normative relations. Because the former duties are non-relational they are incapable of being ingredients of rights under any plausible conception of a right. The latter duties, meanwhile, can be ingredients of rights only under the interest conception. All of the natural rights in these theories are therefore translatable without remainder into the language of relational duties and their correlative claims. I know of no example in this era of a theory which modelled its natural rights on the choice conception; this seems to have been a distinctively modern innovation.

This expulsion of some of the tradition's paradigm instances is not, however, as serious as it might seem. A theory which fails only the first condition can be readily relabelled a natural duty, or natural law, theory. Indeed, this relabelling has the advantage of revealing the deep differences between the classical theories and many modern natural rights theories, differences which are often overlooked because they have been obscured by the common rubric of natural rights. Furthermore, it will not be difficult to broaden the range of our critique so as to include natural duty theories. The two kinds of theory share a commitment to the belief that some deontic category is (metaphysically) natural, as well as an interpretation (provided by the other conditions) of what that commitment means. On the issue of the naturalness of rights/duties they are therefore likely to stand or fall together. In any case, the conceptual landscape has been very different in the modern era. This era continues to see some theories on the classical model.[13] But its more distinctive contribution has been the emergence of a family of theories which have been explicitly associated by their authors with the natural rights tradition and which have tied rights conceptually and functionally to the protection of individual autonomy.[14] While the best-known instances of these theories have been libertarian, there is nothing in the choice conception which commits one to this particular option. Libertarianism as a moral/political theory is based on the contention that natural rights are all liberty-rights (or property rights). But this is a substantive claim about the kinds of rights people have which is in no way entailed by the choice conception. As we have seen, the choice conception can readily accommodate both liberty-rights and claim-rights, and among the latter it can readily accommodate the sorts of welfare rights which are anathema to libertarians. Libertarians can therefore share a conceptual scheme with their opponents—including those who espouse rival theories of natural rights—while defending their distinctive substantive claim about the content of those rights.

In short, in the modern era there is no longer conceptual agreement among natural rights theorists. In the light of this development we could of course have remained agnostic, thus permitting a theory to choose either conception of a right. But theoretical considerations strongly favoured requiring the choice conception, and they have been allowed to prevail. Meanwhile, some of the other conditions also exclude some putative natural rights theories. I know of no such theory which fails the second condition; everyone seems to agree that it is a necessary condition of belonging to the tradition that a theory affirm the existence of natural rights in the merely extensional sense. However, some theories fail the third condition, and not just because they are duty-based. One prominent recent instance of a self-described natural rights theory does indeed accept the interest model of rights, but it fails the third condition not because its basic principles impose

[13] The most noteworthy recent example is Finnis 1980 (see 198–205 for Finnis's endorsement of the interest conception of rights). Note that Finnis offers us the alternative of calling his theory a natural law theory.

[14] See, for instance, Hart 1955; Nozick 1974; Machan 1978.

duties but because they postulate goods.[15] Meanwhile, still other theories fail the fourth condition because they accept a constructivist methodology.[16]

The account constructed here is therefore fussy about what it counts as a natural rights theory. As far as I can see, it does not admit any theory which everyone regards as standing outside the natural rights tradition; but it does deny admission to some theories which some regard as belonging to the tradition. It yields this latter result because three of the conditions which it imposes are not universally accepted within the tradition. On the other hand, all four conditions are at least widely accepted within it; thus none has simply been imposed from the outside. Furthermore, since the tradition is now divided within itself any account which would be universally accepted would be too weak to be of service in our present inquiry. Were we to require, for instance, only that a theory recognize some moral rights which are extensionally natural then the natural rights tradition would no longer offer a distinctive notion of a moral right or a distinctive set of existence conditions for such rights. The demands of our theoretical inquiry have dictated that where the two criteria of adequacy conflict the criterion of theoretical pay-off will dominate that of historical fidelity. Whereas the latter criterion would favour a broader and more permissive account, the former requires a narrower and more restrictive one. Whether everyone will accept the result as an account of the nature of natural rights is unimportant as long as it isolates one promising route to understanding the nature of moral rights.

4.2 WHY RIGHTS CANNOT BE NATURAL

As we have construed it, the natural rights tradition holds that individuals possess certain basic moral rights by virtue of some aspect of their nature and that these rights are conferred on them by rules which are features of the natural order. In order to test this account of the existence conditions for moral rights I aim to reconstruct and develop some criticisms which were directed against it nearly two centuries ago by Jeremy Bentham. Implacable,

---

[15] Finnis 1980.
[16] For example, Rawls 1971; Harman 1980. See the notion of a semi-natural right in Morris 1985.

and often intemperate, opposition to natural rights was one of the constants in Bentham's long career.[17] However, he had many different grounds for this opposition, some of which are much more important than others. We may begin by dividing these grounds into two broad categories: the conceptual and the substantive. The distinction is suggested by one of Bentham's typically pungent dismissals of natural rights: 'The assertion of such rights, absurd in logic, is pernicious in morals.'[18] We will begin with Bentham's reasons for thinking that natural rights are absurd in logic, turning later to their alleged subversion of our moral thinking.

As his accusation of logical absurdity suggests, Bentham believed not just that there are as a matter of fact no natural rights but that there could be no such things; the very idea of a natural right is 'a contradiction in terms'.[19] Furthermore, the incoherence of natural rights does not lie in some further property, such as imprescriptibility, which such rights might be claimed to possess. Natural and imprescriptible rights may be 'nonsense upon stilts' but natural rights *tout court* are still 'simple nonsense'.[20] It was the idea that rights could be natural that Bentham regarded as absurd.

His main argument in support of this allegation runs roughly as follows: (1) there can be no rights without laws; (2) there can be no natural moral laws; therefore, (3) there can be no natural rights. Bentham was led to his first premiss by his assumption of the interest conception of rights: roughly speaking, to have a right in Bentham's view is to be the intended beneficiary of some duty imposed on others. Bentham's conception of a right is thus our conception of a claim; it therefore includes the notion of a (relational) duty. But a duty for Bentham is inconceivable apart from some law that imposes the duty. Thus the argument to the first premiss runs roughly as follows: there can be no rights without duties; there can be no duties without laws; therefore, there can be no rights without laws. This argument is not weakened if we replace the interest conception of a right by the choice conception. It will then run roughly as follows: there can be no rights without both duties and powers; there can be no duties or powers without

---

[17] The story of this opposition may be found in Hart 1982, ch. 3. In ch. 4 Hart offers a valuable reconstruction of Bentham's principal arguments against natural rights. I depart in some respects from Hart's account in my own reconstruction.

[18] Bentham 1843, iii, 221; cf. ii, 497.

[19] Bentham 1952–4, i, 334.   [20] Bentham 1843, ii, 511.

laws; therefore, there can be no rights without laws. Indeed, if anything the argument is stronger for rights (as the choice conception construes them) than it is for mere claims. This is a point to which we will return later.

Bentham was led to the second premiss of his main argument by his concept of a law. According to Bentham a law is (roughly speaking) an expression of the will of a sovereign concerning the conduct of those who are subject to the sovereign's will.[21] The concept of a law thus includes that of a legislator. The argument to the second premiss then runs roughly as follows: there can be no laws without a legislator; natural moral laws have no legislator; therefore, there can be no natural moral laws. This argument is unsatisfactory as it stands because of its reliance on Bentham's volitional account of the existence conditions for laws. As we have seen, more recent positivists have abandoned this account in favour of a more complex and plausible analysis in which the existence of a law is a matter of its validity within a legal system, and the existence of the system as a whole is a matter of its being sustained by conventional social practices of compliance with and acceptance of its rules on the part of those to whom the rules apply. However, the substitution of this improved account does not weaken the argument to the second premiss. It will now run as follows: there can be no laws whose existence is independent of conventional social practices of compliance and acceptance; the existence of natural moral laws must be independent of all such practices; therefore, there can be no natural moral laws.

The main argument, however, is still inadequate because of its confinement of both duties and rights to the rules of municipal legal systems. Bentham plainly regarded these rules as the paradigm cases of laws. But he sometimes went further, declaring that municipal laws were the only real laws.[22] From this he drew the inevitable conclusion, which he was fond of repeating, that legal rights are the only real rights.[23] But this conclusion is a wild exaggeration. As we have seen, there is no obstacle in the way of recognizing non-legal rule systems as sources of both duties and rights. In his more careful moments Bentham was aware of this fact, for he regarded the rules of a society's positive morality as

[21] For the full story see Bentham 1970(*b*).
[22] As in Bentham 1977, 7 ff.
[23] See Bentham 1843, ii, 500, 523; iii, 218–21; Bentham 1952–4, i, 324.

sources of both duties and rights.[24] Bentham therefore had no good reason to confine duties and rights to the rules of municipal legal systems. But the argument is once again not weakened by deleting this restriction. The argument to the second premiss will now run as follows: there can be no rules capable of conferring rights whose existence is independent of conventional social practices of compliance and acceptance; the existence of natural moral rules must be independent of all such practices; therefore, there can be no natural moral rules. And the main argument will now run: (1) there can be no rights without rules; (2) there can be no natural moral rules; therefore, (3) there can be no natural rights.

The first premiss of this argument is not at issue between natural rights theorists and their opponents. Those who affirm the existence of natural rights also affirm the existence of natural rules (or laws) which confer those rights. But the second premiss is at the heart of the issue. Bentham's argument to this premiss makes it easy to locate the point of contention. Natural rights theorists agree that the existence of natural moral rules (or laws) is not dependent on any actual conventional practices; the natural moral law is not a conventional rule system. Thus the keystone of Bentham's case against the intelligibility of natural rights is the thesis that there are no non-conventional rule systems capable of creating rights. Bentham provided little or no explicit argument in support of this thesis; it appeared to him to be so obviously true as merely to require stating. But the thesis appeared so obvious to Bentham because it was so strongly (though also negatively) supported by a positivist account of the existence conditions for conventional rule systems. In this sense the needed support for the thesis is to be found in Bentham's life work.

Before turning to the question of how the thesis might be supported, we need to be clear about its meaning. It does not claim that there are no non-conventional rule systems. Thus it does not claim that the rules of arithmetic or natural deduction either fail to form a system or are merely conventional (or both). Instead, the thesis claims that there are no non-conventional rule systems capable of conferring rights, thus capable of conferring (or imposing) the various Hohfeldian positions which are the ingredients of rights. Nor does the thesis claim that there are no moral rights.

[24] Bentham 1843, viii, 247; 1970(*a*), 34–5; 1977, 496.

Instead, it claims, or rather implies, that there are no natural moral rights. Finally, the thesis does not claim that there are no natural laws. It is, as it must be, quite compatible with the existence of natural causal laws. Instead, it claims that there are no natural *moral* laws—no natural laws, that is, which are capable of conferring moral rights.

Bentham's defence of the thesis consists of an implicit challenge. If some rule system is claimed to exist then it must be possible to provide its existence conditions. We know what the existence conditions are for conventional rule systems. But a system of natural moral rules is not conventional. We should therefore not expect the same account of its existence conditions. But we should expect some analogue to this account. What could this be? Bentham's thesis amounts to the claim that no such account can be given, precisely because the rules in this case are not conventional.

The challenge can be sharpened by attending to the notion of a right. On the choice conception a right is a bundle of Hohfeldian positions. These positions divide into two categories: first-order positions which determine what it is permissible for individuals to do and second-order positions which determine what it is possible for them to do. Let us continue to treat duties as the basic first-order positions and powers as the basic second-order positions. Since all of the other ingredients of rights can be constructed out of these two, an account can make sense of natural rights if it can make sense of both natural duties and natural powers.

Consider duties first. Duties, of whatever variety, necessarily have normative force; that is, to have a duty imposed by some particular rule system is necessarily to have a reason for acting relative to that system. The existence conditions for any particular variety of duties must therefore be capable of explaining how those duties come to have normative force, thus how they come to provide practical reasons. Where a duty is conventional the needed explanation is provided by the complex social practices which sustain it. In the case of a legal duty, for example, once we know that a given legal system is in force, that the rule imposing the duty is valid within that system, and that this rule applies to a particular individual—once we know all of these things we need nothing more in order to explain why that individual has a legal reason to do what the duty requires. The same sort of account will equally explain the normative force of non-legal, but still conventional, duties. In all such cases the rules of the system impose duties with normative force ultimately because those rules are regarded as doing so—that is, are complied with and/or accepted—by those to whom they apply. The normative force of conventional duties is thus a social artifice.

There seems to be no analogous account available to us in the case of natural duties. These duties must of course have moral force. But in order to explain their moral force we cannot appeal to any anchoring notions of authority, compliance, or acceptance. What makes natural moral rules natural is precisely the fact that they issue from no authority and are valid regardless of whether they are complied with or accepted. But in that case how could we ever begin to explain how they can create moral reasons for action, indeed how they can be normative at all? The naturalness of natural laws seems to preclude their being normative, and thus the naturalness of natural duties seems to preclude their being duties—or so Bentham seems to have thought.

There is a resource available here to the would-be natural rights (or natural duty) theorist. The problem of the normative force of natural laws would be solved if they could be shown to have issued from some legislator. Since no human legislator could serve this function it is tempting to invoke a divine analogue. Appeals to some deity as the authoritative source of natural laws were a common theme in classical natural duty theories. Indeed, it is striking just how many natural law theorists agreed with Bentham that a law cannot be binding—cannot impose duties—unless it has been commanded by some sovereign. Thus, for instance, Locke held that 'what duty is, cannot be understood without a law; nor a law be known or supposed without a lawmaker, or without reward and punishment'.[25] Since Locke believed in the existence of a natural moral law he took the inevitable step:

. . . it is pretty clear that all the requisites of a law are found in natural law. For, in the first place, it is the decree of a superior will, wherein the formal cause of a law appears to consist. . . . Secondly, it lays down what is and is not to be done, which is the proper function of a law. Thirdly, it binds men, for it contains in itself all that is requisite to create an obligation.[26]

---

[25] Locke 1894, i, 76; cf. Locke 1954, 181–3, 187; Pufendorf 1934, 89.
[26] Locke 1954, 111–13; cf. 181–3, 189; Locke 1894, i, 69–70, 474–5. Pufendorf takes the same step: Pufendorf 1934, 215, 217.

Not all natural law theorists have shared the view that natural laws can have normative force only if they are also divine laws.[27] Those who choose not to pursue the theological route will need to find some other way of responding to Bentham's challenge. Meanwhile, let us consider the extent to which the invocation of a deity solves the problem. It introduces, of course, evidential uncertainties. It is a central claim of natural law theories, whether duty-based or right-based, that the content of the law of nature is discoverable by the use of our natural faculties. Thus if a deity is to play this crucial legislative role within such a theory then it must be possible, at least in principle, to establish the existence and nature of this deity by rational means. On the strategy under examination, therefore, a natural law theory presupposes a natural theology.[28] The difficulties involved in defending such a theology are notorious, but unless they can be overcome the theological strategy will be ineffectual.[29] Let us therefore suppose that they can be overcome. Could a theological natural duty theory then provide existence conditions for natural duties?

Sadly not. The existence of a divine legislator might indeed explain the existence of a rule system capable of imposing duties.[30] But these duties will have no moral force. Rules created by a deity and imposed upon a community of subjects will merely constitute another conventional rule system—the legal system of another community. Conventional rule systems are capable of imposing duties with normative force but not duties with moral force. What holds for human positive law will hold as well for divine positive law (if there is any): in order to impose duties with moral force it must satisfy some further, non-conventional, moral conditions. These conditions might be sought in the good nature of the deity or, as with Locke, in the duty of the subjects to obey.[31] But on the former option some principles of the good are basic to the theory, while on the latter some moral duties exist independently of the natural law. In either case the enterprise of a natural duty theory has been abandoned.

[27] For an illustrious exception see Grotius 1925, 38–45.
[28] As Locke recognized: Locke 1954, 151 ff.
[29] For Bentham's scepticism concerning this project see Bentham 1952–4, i, 334 n.
[30] Though not if this system is simply imposed by dint of superior force; it must be accepted by at least some of those subject to it.
[31] Locke 1954, 151 ff., 181–3.

The theological strategy thus appears to be a dead end. And there seems no alternative way of explaining how natural rules can impose duties with moral force. Natural duty theories thus appear to be unable to provide existence conditions for the duties which they affirm. Since rights include duties, this conclusion strikes against natural rights theories as well. But since rights include more than duties, the burden to be discharged by natural rights theories is even greater. Consider now the other principal ingredient in rights, namely powers. Suppose that we had in hand some account of the existence conditions for natural duties. This would be sufficient to establish the coherence of natural duty theories but not natural rights theories, since it remains possible that individuals lack the power to manipulate these natural duties. Indeed, the fact that the duties would be natural strongly suggests that they would lack this power. After all, a natural duty would be a natural fact created by a natural law. The ability to alter such facts at will, by exercising some moral power, would be very odd indeed; it would seem to constitute some magical control over nature. Natural rights theories thus suffer from a special impediment which is not shared by natural duty theories; they must explain the nature and source of this mysterious power, again without reliance on the conventional practices which sustain the exercise of conventional powers. Because Bentham accepted the interest conception of a right his original argument was aimed only at what we are calling natural duty theories; it did not trouble itself over the possibility of natural powers. When I reconstructed the argument earlier so as to employ the choice conception I noted that it was thereby strengthened. We can now see why this was so.

Bentham's argument is a sceptical argument. To the extent that it is successful, it serves to cast doubt on the enterprise of providing existence conditions for natural rights. It does so by contrasting these rights unfavourably with conventional rights. We know how to provide existence conditions for the latter but it is obscure just how this story could ever be extrapolated to cover the former. The argument does not demonstrate that there are no existence conditions for natural rights or that rights cannot be natural (though Bentham himself had no doubts on these matters). But it does construct a case for the natural rights theorist to answer. Perhaps there is an answer, though no one seems to know what it is. In advance of supplying it, however, the natural rights tradition does not move us any closer to finding existence conditions for moral rights. By contrast with conventional rights, moral rights are mysterious. The natural rights tradition tells us that just as conventional rights are the products of conventional rule systems, moral rights are the products of a natural rule system. But in the end the notion of a system of natural rules capable of conferring rights turns out to be every bit as mysterious as the notion of a moral right with which we began.

Although a sceptical argument like Bentham's cannot show conclusively that the very idea of a natural right is incoherent, it would be strengthened if some alternative explanation could be given for the persistent belief in the existence of natural rights. Bentham also thought he had such an explanation.[32] Suppose that the legal system governing us has denied us some legal right which we believe we ought to have and that we wish to make a case for being accorded that right. Since we believe that we ought to have the right then we believe that in an ideal legal system we would have it. It is then but a short and tempting step to claiming that an ideal legal system exists in which we do have the right, and then to saying that the right conferred on us by this ideal system constitutes our case for having the same right conferred on us by our actual system. The mechanism at work here is projection: as a corrective to the deontic imperfections of the actual world we invent a deontically perfect world in which individuals possess just those rights which, morally speaking, they ought to have, and then we treat this invention as though it were real. This projective mechanism is an obvious route to a natural rights framework. Indeed it is such an obvious route that it intensifies the force of Bentham's challenge. Unless natural rights theorists are able to supply existence conditions for natural moral laws their belief in such things, like belief in a deity, will continue to seem mere wishful thinking.

I noted earlier that Bentham's objections to natural rights divided into two categories: the conceptual and the substantive. Having explored his reasons for thinking such rights 'absurd in logic' I turn now to the allegation that they are 'pernicious in morals'. Of course, if the conceptual challenge cannot be met then there can be no natural rights and the question of their role in our moral

---

[32] Bentham 1843, ii, 501; iii, 221.

thinking does not arise. I shall therefore suppose that the conceptual challenge can be met, thus that a coherent natural rights theory can be constructed. Our question now is whether such a theory could do any useful moral work.

As in the case of his conceptual arguments, some of Bentham's substantive objections to natural rights are more important than others. Bentham believed that any attempt to derive moral or political conclusions from premisses about individual rights was mischievous, but some of the alleged mischiefs stem from features which are not essential to natural rights. We must remember that for Bentham appeals to such rights in the service of political causes were more than abstract possibilities. Twice during his lifetime a major revolution was defended on just these grounds. The documents which occasioned Bentham's fulminations against natural rights were the manifestos issued in defence of these revolutions. In the case of the Declaration of Independence Bentham eventually became reconciled to, indeed a champion of, the resulting United States of America—though he continued to believe that it had been built on bad arguments.[33] But when he responded to the French Declaration of the Rights of Man and the Citizen he was animated by horror at the violence that had been unleashed by the revolution in France. Keeping this violence in mind will help us to understand why Bentham came to regard natural rights as the language of anarchists and terrorists.

Even in his calmer and more reflective moments, however, Bentham regarded appeals to natural rights as dangerous. In support of this belief he made an important distinction. The claim that natural rights impose moral constraints on conventional social arrangements may be interpreted in two quite different ways. On the first interpretation these rights limit what governments may do—what it is permissible for them to do. On the second interpretation they limit what governments can do—what it is possible for them to do. The two interpretations diverge in their account of what has happened, or failed to happen, when a government has ostensibly enacted a law which infringes some natural right. On the first interpretation, which results from combining a natural rights theory with a positivist theory of law, the government has enacted an unjust law. On the second

[33] See Hart 1982, ch. 3.

interpretation, which results from combining a natural rights theory with a natural law theory of law, the government has enacted no law at all.

Consider first the coalition of natural rights and legal positivism. On this view some set of rights serves as a standard of the justice of laws but not as one of their existence conditions. Bentham agreed that some moral standard was necessary for evaluating the law but rejected the idea that it should be formulated in terms of rights. Some of his grounds for this rejection stemmed from certain features of the rights declarations with which he was familiar.[34] It was common for the rights in these declarations both to be characterized in a very broad and sweeping manner and to be treated as indefeasible or absolute. One standard example was the right to liberty. As Bentham never tired of pointing out, every law which imposes a duty necessarily limits the liberty of those to whom it applies. It follows that if everyone possesses a right to liberty which is both unlimited and indefeasible a government may impose no duties at all, which is to say that it may not govern. Since natural rights are neither broad nor indefeasible by nature, this problem does not stem from their nature. On the other hand, it is also not easily remedied. If we retain the indefeasibility of rights but characterize them more carefully then they may indeed leave spaces which a government may fill with legislation. But the absolute constraints that they impose are still likely to be too rigid to adapt readily to fluctuating social and political circumstances. On the other hand, if we treat rights as defeasible then we must determine which considerations are sufficient to defeat them. It is possible, to be sure, that nothing can override a right except another right, but in that case a set of basic natural rights will have to be glossed with a set of priority rules for resolving conflicts. Meanwhile, if any other consideration is allowed to justify the infringement of a right then something other than rights is being treated as morally basic.

These difficulties will of course equally afflict the account which combines a natural rights theory of morality and a natural law theory of law. But this combination also produces its own special mischief.[35] Bentham believed, as do most of us, that the fact that a law is unjust does not suffice by itself to justify refusing to comply with it. In deciding whether to comply with an unjust law we must

---

[34] Bentham 1843, ii, 496 ff.
[35] Ibid. ii, 494–5, 500, 511; cf. Hart 1982, 81–2.

also take into account the alternative courses of action open to us, the likely consequences of our disobedience, and so on. There is therefore no direct inference from the injustice of a law to the justifiability of defying it. Combining a natural rights moral theory with a natural law legal theory, however, threatens to license just this inference. On such a view an ostensible law which infringes a natural right is in fact no law at all. Non-compliance with the 'law' is therefore not truly illegal. But then there is no disobedience which needs to be justified. A natural rights morality plus a natural law legal theory thus appear to provide a simple and direct justification for disobeying, or rather disregarding, unjust 'laws'. It was chiefly for this reason that Bentham regarded natural rights as anarchical.

However, we have not yet reached Bentham's deepest objection to appeals to natural rights in political argument, an objection which applies with equal force regardless of the legal theory to which such rights are harnessed. The objection is best stated as a contrast between appeals to rights and appeals to Bentham's preferred standard, namely utility. The question whether some particular law promotes the general welfare is an empirical one and is therefore answerable, at least in principle. Utility therefore can provide a determinate standard of moral assessment. However, the question whether some particular law infringes natural rights is not an empirical one and is therefore unanswerable, even in principle. Natural rights therefore cannot provide a determinate standard of moral assessment.

In this simple form the argument is defective in a number of respects. Its most important defect for our purposes is that it constructs a misleading analogy between natural rights and utilitarian moral theories. These two types of theory may both be partitioned into base and superstructure. The moral base of a theory, as we know already, contains its basic principles; its superstructure thus contains all of the ingredients (principles, rules, policies, etc.) that are derived from those basic principles. The moral base of a natural rights theory consists of some set of rights principles. The moral base of a utilitarian theory consists of (some version of) the principle of utility. We must now distinguish between arguments to and arguments from a theory's moral base. Arguments to basic principles are non-moral arguments attempting to show that the principles are correct or that acceptance of them is reasonable. Both rights principles and the principle of utility must be supported by some such arguments if we are to have any reason to acknowledge them as a standard of moral assessment. The question whether we should accept some set of rights principles does not look like an empirical one, but neither does the question whether we should accept the principle of utility. Arguments from basic principles are moral arguments attempting to show that conclusions about particular cases are correct or that acceptance of them is reasonable. Both rights principles and the principle of utility must support some such arguments if we are to be able to employ them as a standard of moral assessment. The question whether a particular law promotes the general welfare does look like an empirical one, but so does the question whether the law infringes some specified right. Appeals to rights as a standard of moral assessment thus appear to be analogous in every respect to appeals to utility.

Bentham's argument that appeals to natural rights, unlike appeals to utility, are inherently undecidable does indeed confuse the separate issues of arguments to and arguments from basic principles. But the main point of his objection can be directed specifically at the former issue. Bentham could concede that once a determinate set of basic rights has been established then the question whether a particular law infringes any of these rights is decidable. However, having conceded this he could continue to maintain that arguments to a determinate set of basic rights are inherently undecidable. And Bentham certainly believed that such arguments *are* undecidable, thus that there is no way of showing *which* set of rights should serve as the basic standard of moral assessment.[36] The selection of any particular set of rights is entirely arbitrary because rationally unsupportable; the language of natural rights is 'from the beginning to the end so much flat assertion'.[37] Bentham is of course committed to the additional claim that the selection of a particular version of the principle of utility is not arbitrary because rationally supportable. We cannot decide here whether this claim is justified. But we can ask whether there are any special difficulties involved in arguing to basic principles of natural rights.

It will help here to remind ourselves of the demand we are making of a theory of rights. Our problem is the population

[36] See, for example, the polemic in Bentham 1970(*a*), 21–5 n.
[37] Bentham 1952–4, i, 335.

explosion of rights claims. This explosion can be controlled only by a standard of authenticity for rights. We are looking to a theory of rights to supply such a standard. A rights theory will authenticate all and only those rights that are derivable from its basic principles; those principles thus serve as its ultimate control over the proliferation of rights claims. Two sorts of control are possible. Rights claims are externally controlled if they are tested by means of basic principles which are not themselves rights principles; in that case rights constitute a derivative but not a basic moral category. Rights claims are internally controlled if they are tested by means of basic rights principles; in that case rights constitute both a derivative and a basic moral category. Because it treats rights as morally basic, a natural rights theory is committed to the internal control of rights claims. It is obvious that this strategy can succeed only if there is in turn some control over the proliferation of basic rights principles, thus some standard of authenticity for basic rights. Since these rights are basic they cannot be authenticated by any further, deeper moral principles. A natural rights theory must therefore look beyond morality for a standard of authenticity for its basic rights.

A set of basic rights will supply a determinate test of authenticity for derivative rights only if it is itself determinate. Recall our earlier distinction of the three dimensions of a right: its content, scope, and strength. The content of a right is what it is a right to do or to have done; this is given by the content of its core liberty or claim. The scope of a right has two ingredients: the subjects of the right and its objects. The subjects of a right are those who hold it; this is given by the holders of the various Hohfeldian advantages (liberties, claims, powers, immunities) which are ingredients of the right. The objects of a right are those against whom it is held; this is given by the bearers of the various Hohfeldian disadvantages (chiefly duties and disabilities) which are correlated with ingredients of the right. It follows that the content and scope of a right have been completely specified only when the content and scope of its several ingredients, both core and peripheral, have been completely specified. Finally, the strength of a right is its ability to override, or susceptibility to being overridden by, competing moral considerations. The strength of a right has been completely specified when its weight has been given relative to every sort of consideration with which it might compete.

A right is fully determinate only when its content, scope, and strength have been fully specified. A set of rights is fully determinate only when all of its members are fully determinate. Only a set of fully determinate basic rights can provide a fully determinate standard of authenticity for rights. The best case, therefore, is that a theory of rights will justify some unique set of fully determinate rights. Doubtless this ideal of perfect determinacy, even if attainable in principle, is unattainable in practice. We should therefore settle for any theory which can justify some set of reasonably determinate rights. If a natural rights theory is to rebut Bentham's charge of arbitrariness it must be capable of at least this much. It must therefore be able to argue to some set of rights which are fairly determinate in their content, scope, and strength, and it must also tell us how to determine these dimensions more precisely when it is of practical urgency to do so.

A natural rights theory must look to nature to determine its set of basic rights. An argument from nature will move from empirical premisses describing natural facts to rights principles. The ambition of a natural rights theory is to select that set of basic rights which is most congruent with the natural facts; this is part of what is meant by calling rights natural. But with which natural facts are rights to be congruent? To begin with, how do we decide *whose* nature is relevant? The answer within the natural rights tradition has usually been *human* nature, but how do we know that this is the proper point of departure for an argument from nature? Starting with human nature will lead us to human rights, thus fixing the scope of basic rights. But why is it not simply arbitrary to assume this criterion for rights? How could we defend one starting-point rather than another? And if we cannot, how could an argument from nature even determine the scope of basic rights? Furthermore, even if we do confine attention to our own nature, which aspects of our nature are the relevant ones? We are beings capable of choice—do we therefore have liberty-rights? We are also beings capable of being injured by others—do we therefore have claim-rights not to be harmed? We are also beings who need the support and assistance of others—do we therefore have claim-rights to be given such assistance? If we lack any of these rights, why do we lack them? If we have them all then how can our nature determine which is to take precedence when they conflict? How, in general, can we distinguish between the relevant and the irrelevant aspects of our nature without presupposing a particular outcome for the argument? The problem here is not that arguments from nature are invalid because they commit some naturalistic fallacy. The problem is that too many such arguments seem to be valid, and that there seems no way to arbitrate among them by further appeals to the facts. But if this is so then nature, even our nature, underdetermines selection of a set of basic rights and thus provides no effective control over the proliferation of basic rights principles.

There is a general problem about arguments from nature: they always threaten to be either inconclusive or circular.[38] This is a problem confronting any moral theory which accepts a realist methodology. But the problem is particularly acute for arguments which run directly from nature to basic principles of rights. Some moral categories, such as the good, seem to be more naturalistic than others. Thus one can imagine successful arguments which run directly from nature to basic principles of the good—that is, a plausible naturalistic value theory. But deontic categories seem the least naturalistic, by virtue of their origins in conventional rule systems. Thus it is harder to imagine successful arguments which run directly from nature to basic principles of duty—that is, a plausible natural duty theory. And rights seem the least naturalistic of all deontic categories, by virtue of their complex structure and their inclusion of second-order Hohfeldian elements. Thus it is hardest to imagine successful arguments which run directly from nature to basic principles of rights—that is, a plausible natural rights theory. But that means that even within the class of theories which share a realist methodology natural rights theories seem the least likely to succeed.

4.3 THEORETICAL ALTERNATIVES

Bentham's conceptual and substantive arguments against natural rights are both sceptical challenges. Both are therefore vulnerable to rebuttal. But I know of no natural rights theory which has ever provided such rebuttals, and until they are forthcoming we are entitled to doubt that any such theory can supply existence conditions for moral rights. Furthermore, the two challenges are interlocked. Although existence conditions for moral rights will not

---

[38] An excellent critique of such arguments can be found in de Sousa 1985.

by themselves tell us which rights there are, they will at least give us a procedure for discovering this. Thus if we had existence conditions for natural rights we might also be able to establish a sufficiently determinate set of such rights. The seeming incoherence of natural rights is thus their main impediment; their seeming indeterminacy is a subsidiary problem.

If the sceptical challenges cannot be successfully rebutted then the result is paradoxical: the very theories which have taken rights most seriously are incapable of showing that rights should be taken seriously. Thus one route to a set of existence conditions for moral rights is closed off. Others, however, remain open. No attempt will be made here to enumerate all of the remaining candidates, but two of them merit special attention.

As I have characterized it, a natural rights theory satisfies four conditions: (1) it contains some rights (on the model of protected choices), (2) at least some of these rights have a natural criterion, (3) the theory treats those rights with a natural criterion as morally basic and treats nothing else as morally basic, and (4) it claims that these basic rights are objective and thus accepts a realist methodology. The first condition must be satisfied by any rights theory, and the second must be satisfied by any theory which treats some of its rights as natural in the merely extensional sense. But the outcome of Bentham's arguments is the conclusion that no theory can additionally satisfy both the third and fourth condition. If a theory elects to treat rights as structurally basic then it cannot also treat them as objective moral facts, and if it elects to treat rights as objective then it cannot also treat them as basic. Bentham's arguments are thus, in effect, impossibility arguments: no theory capable of supplying existence conditions for moral rights can satisfy all four conditions. Furthermore, the arguments suffice to reject not only natural rights theories but natural duty theories as well (including those theories which accept the conception of rights as protected interests). The defect in all such theories is that no deontic category can be both basic and objective.

This result, however, suggests an obvious remedy: retain the first two conditions (thus natural rights in the extensional sense) and discard at least one of the others. Rights theories in both the contractarian and consequentialist traditions take up this option; they disagree, however, on which condition is to be discarded. Contractarian theories are committed to discarding the fourth condition. Thus they may regard rights as morally basic, but they also regard them as subjective. Consequentialist theories, by contrast, are committed to discarding the third condition. Thus they may regard rights as objective, but they also regard them as morally derivative. These two options are importantly different. The first may leave intact the moral content of a natural rights theory while rejecting its method. The second may leave intact the method of a natural rights theory while rejecting its content. Because neither option attempts to satisfy all of the conditions which jointly define a natural rights theory, both are immune to Bentham's arguments. Both therefore merit further investigation.

The story told by natural rights theories about the existence conditions for moral rights is so appealing in part because it is so simple. It would be very satisfying indeed to believe in the existence of a Platonic rule system capable of conferring such rights. If we reluctantly abandon this belief, along with the companion belief in a deity, we need not fall into nihilism about moral rights. The stories told by the alternative theories are likely to be more complicated, but they may also, perhaps for this very reason, be more compelling.

5

Contractarian Rights

ALL contractarian theories agree that the principles of morality, or at least the principles of justice, must be grounded in some procedure of collective choice. Beyond this broad area of agreement, however, different theories in the tradition are capable of diverging both in their methodology and in their content. The methodological divergence results from competing interpretations of the appropriate collective choice procedure. Because this issue cuts to the heart of the contractarian tradition, attention to it is best postponed until later in the chapter. Meanwhile, on the side of content a contractarian theory may exhibit any of four possible moral structures: (1) it may contain no rights whatever; (2) it may contain rights but treat none of them as basic; (3) it may treat rights as basic along with some other items; (4) it may treat nothing but rights as basic. In what follows I shall ignore the first two options. Since contractarianism is standardly a theory of justice, the likelihood that a particular theory would contain no rights at all might seem very remote.[1] It will seem less remote, however, if we remind ourselves that we have accepted the model of rights as protected choices. Thus a theory might seek to ground principles of justice, and also connect justice and rights, while interpreting rights as protected interests. Such a theory would be the contractarian analogue of a natural duty theory. Since the fate of contractarian rights theories in no way depends on the resolution of this conceptual issue, it will be convenient to proceed as though all such theories accept the choice model. (The critique to be developed later will in any case apply equally to those which prefer the interest model.)

The second of the four possible structures, on the other hand, is a

---

[1] It might be thought that the utilitarian morality which is derived from a contractarian methodology in Harsanyi 1982 and 1985 is an obvious instance. But Harsanyi favours his version of rule-utilitarianism in part precisely because it makes room for rights.

genuinely important alternative. However, because it has been taken up more often by consequentialists than by contractarians, we will avoid needless duplication if we postpone consideration of it until the next chapter. These exclusions leave us, then, with two interesting cases: a mixed theory and a right-based theory. Although I shall focus primarily on the latter there seems no good reason to exclude the former from the scope of our investigation.

The theories which we will consider all agree, therefore, that some rights are morally basic and that they are to be grounded in some collective choice procedure. In virtue of this constructivist methodology, such theories treat their basic rights not as natural facts which we are capable of discovering but as artefacts which we are capable of inventing. We must, however, be careful here. We have seen already that conventional rights are artefacts, created and sustained by complex social practices. The contractarian must not be interpreted as claiming that moral rights are merely conventional rights. Since moral rights necessarily have moral force, and since conventional rights may lack such force, this claim could not be true. In grounding moral rights in some agreement or bargain the contractarian is not deriving them from any actual social practice. Instead, the rights which the contractarian treats as morally basic are those which would result from some specified collective choice procedure. Whether any such procedure has ever in fact yielded any such rights is irrelevant; basic rights are the products not of any actual agreement but of some possible or hypothetical agreement.

While contractarians need not hold that moral rights are conventional, they must hold that they are subjective. Indeed, most contractarians hold the stronger view that all values are subjective.[2] It is easy to see why they would be led to this view. Suppose that there are some objective, but non-deontic, values—for example, some natural goods. If the existence of these goods is once admitted then it will be difficult to deny them some substantive role in the grounding of rights. But assigning these goods any such role would be incompatible with grounding rights solely in a collective choice procedure. Subjectivism about rights, or about any deontic values, appears to be incompatible with objectivism about goods, or about any non-deontic values. Contractarians thus appear to be committed to the view that there are no objective values.

[2] See, for example, Hobbes 1968, ch. 6; Mackie 1977, ch. 1; Gauthier 1986, ch. 2.

If this is so, then a contractarian theory of rights will be no more plausible than its underlying subjectivism about values. A constructivist methodology is inappropriate in any domain in which a more objective confirmation procedure is available—which is why there are no contractarian theories of physics or geography.

It might be thought, then, that the onus lies on contractarians to show that their methodology is appropriate in the moral domain, and that they can do this only by successfully defending their value subjectivism. Although some contractarians have attempted to meet this demand, it would be unfair to impose it as a condition of adequacy for a contractarian rights theory. It is not obvious, and not easy to show, either that values are subjective or that they are objective. Contractarian moral theories begin with the former thesis, while their methodological rivals begin with the latter. Because each thesis is difficult to defend *ab initio*, each functions in effect as a working assumption for the theory in question. The assumption underlying a particular theoretical enterprise is then confirmed to the extent that the enterprise subsequently goes well and disconfirmed to the extent that it goes badly. Since our enterprise consists in finding existence conditions for moral rights, we may take success or failure at this task as confirmation or disconfirmation of a working assumption about the ontological status of rights. As we have seen, natural rights theories operate with the assumption that rights are objective. Their failure to yield existence conditions for rights thus counts to some extent against this assumption. (It does not count decisively against it, since rights might be objective without being basic.) If contractarians succeed in providing existence conditions for rights then their success will count heavily in favour of their subjectivism about values, while if they do not then their failure will count against it. There is no initial burden of proof so far as views about the status of values are concerned; the view we should accept is the one which turns out to work best.

The main aim of this chapter is to determine whether the contractarian tradition can yield existence conditions for moral rights. However, before turning to this task there is a piece of unfinished business to be dealt with. Thus far we have found the notion of a moral right essentially mysterious. Our success in constructing existence conditions for conventional rights did nothing to dispel the mystery, since it yielded no explanation of how rights could have moral force. When we turned for such an explanation to the natural rights tradition the story we were then told, that the moral force of rights is generated by a system of natural moral rules, turned out to be an even deeper mystery. Thus we still lack an account of what makes a right a moral right. Since the conceptual resource which we need can be found in the contractarian tradition, this seems the appropriate time to develop it.

5.1 THE CONCEPT OF A MORAL RIGHT

Although the following account of the nature of moral rights is implicit in, or presupposed by, all contractarian rights theories, its best explicit presentation has been by a consequentialist. (As will be clear later, the account is the common property of contractarians and consequentialists alike.) Whereas Bentham served in the preceding chapter to cast doubt on the very idea of a moral right, John Stuart Mill will help us to retrieve that idea. In *Utilitarianism*, as everyone knows, Mill set out to defend the view that 'actions are right in proportion as they tend to promote happiness, wrong as they tend to produce the reverse of happiness'.[3] Having done his best with this task in the first four sections of the essay, he turned in the last section to the implications of his moral theory for questions of justice. The concept of justice, as Mill understood it, is inextricably connected with the concepts of duty (or obligation) and of rights. Since justice, in the sense which interested Mill, is a moral notion, these others must also be moral notions. Mill's enterprise therefore required him to make sense both of moral duties and of moral rights, all within a consequentialist framework.

Mill's consequentialism plays no part in his conceptual analysis of justice; thus we may safely ignore it here. Mill approached the concept of justice in two stages, first constructing the notion of a moral duty and then using this in turn to isolate the notion of a moral right. Since duties are simpler than rights, by virtue of being one of their ingredients, we will do well to adopt Mill's two-stage strategy. The following passage then gives us the essentials of Mill's analysis of moral duties:

We do not call anything wrong, unless we mean to imply that a person

[3] Mill 1969, 210.

ought to be punished in some way or other for doing it; if not by law, by the opinion of his fellow creatures; if not by opinion, by the reproaches of his own conscience. This seems the real turning point of the distinction between morality and simple expediency. It is a part of the notion of Duty in every one of its forms, that a person may rightfully be compelled to fulfil it. Duty is a thing which may be *exacted* from a person, as one exacts a debt. Unless we think that it might be exacted from him, we do not call it his duty. Reasons of prudence, or the interest of other people, may militate against actually exacting it; but the person himself, it is clearly understood, would not be entitled to complain. There are other things, on the contrary, which we wish that people should do, which we like or admire them for doing, perhaps dislike or despise them for not doing, but yet admit that they are not bound to do; it is not a case of moral obligation; we do not blame them, that is, we do not think that they are proper objects of punishment . . . I think there is no doubt that this distinction lies at the bottom of the notions of right and wrong; that we call any conduct wrong, or employ, instead, some other term of dislike or disparagement, according as we think that the person ought, or ought not, be punished for it; and we say that it would be right to do so and so, or merely that it would be desirable or laudable, according as we would wish to see the person whom it concerns, compelled, or only persuaded and exhorted, to act in that manner.[4]

Mill's views, as outlined in this passage, contain some idio-syncrasies which we are not compelled to adopt. One of them is the idea that the language of morals is entirely the language of duty, or obligation, or right and wrong. This has the curious implication that no assessment of conduct as desirable or laudable, but not obligatory, is a moral assessment. Modern moral philosophy draws the conceptual boundaries differently, so that what Mill took to be 'the distinction between morality and simple expediency' we would now regard as the distinction between the morally obligatory and the morally valuable, or between deontic and non-deontic moral assessments. But this is no matter; Mill's analysis of moral duties is easily detached from his views about the extent of the moral domain.

A more central feature of Mill's account is its reliance on sanctions. Not to put too fine a point on it, Mill's view is that I have a moral duty to do something if I ought to be threatened with some form of punishment in the event of my failure to do it. It is true, of course, that Mill does not confine his attention to legal sanctions;

[4] Ibid. 246.

thus the punishment in question may be prescribed by law but it may also be inflicted by my fellow creatures or by my own conscience. But why was Mill so fixated on sanctions in the first place? The answer, I think, lies in certain assumptions which he was making about the nature of conventional duties. Recall our own earlier treatment of such duties. While conventional duties are imposed by conventional rule systems, some of the rules of such systems may not have the function of formulating deontic constraints; they may instead set guidelines or standards for desirable or laudable conduct. How, then, are we to distinguish between those rules which dictate what we must do and those which indicate what we should do or what it would be good for us to do? Our earlier answer to this question pointed to two common characteristics of rules which impose duties: their urgency (the force of 'must' v. 'should') and the polarity of the reinforcing social pressure behind them (the punishment of failure v. the rewarding of success). In the above passage Mill is obviously trying to draw the same sort of boundary in the same sort of way; thus blame and punishment are aligned with the obligatory, while admiration and exhortation characterize the desirable or laudable. Then, having connected conventional duties with sanctions, Mill does the same for moral duties.

This linkage is understandable, but none the less mistaken. It is true that the use of some negative reinforcement is a standard, and generally reliable, sign of a conventional duty; but it is not true that it is a necessary feature of a duty. Duties with no reinforcing sanctions are not even mere logical possibilities; some of the duties in most, if not all, legal systems fall into this category.[5] The reason that there is conceptual room for sanctionless duties is that the imposition of a sanction is not the only means available for conveying the message that some particular sort of conduct is demanded rather than merely recommended or welcomed. (Another means is to say just that.) Mill appears to have inherited from Bentham the view that every conventional duty, whether legal or non-legal, must include some penalty to be inflicted in the event of failure.[6] Where the duty is imposed by a legal system the sanction

[5] See Raz 1975, 158.

[6] The connection which Bentham affirmed between legal duties and legal sanctions is examined in Hacker 1973 and Hart 1982, ch. 6. Bentham's full classification of sanctions, legal and non-legal, may be found in 1970(*a*), ch. 3.

will be prescribed and administered by law; where it is imposed by the rules of a conventional morality the sanction will be administered by the opinion of one's fellow creatures or (if the rule has been internalized) by the reproaches of one's own conscience. But for Bentham and Mill alike, in any case in which there is no sanction there is also no duty.

Mill's reliance on sanctions in his account of moral duties is thus explainable as his way of isolating those conventional rules whose function is to impose duties. In that case sanctions are a needless, indeed distracting, ingredient of his analysis. What can be salvaged from that analysis is the linkage between moral duties and conventional duties. We know that the existence conditions for the latter cannot serve as the existence conditions for the former. Thus we cannot say that I have a moral duty just in case I have a conventional duty with the same content. What, then, could the relationship be between a moral duty and the corresponding conventional duty? The answer which Mill gives to this question has been staring us in the face all along: a moral duty is not just a conventional duty but it is a morally justified conventional duty. We know that the existence of a conventional duty depends on the existence of the conventional rules which impose the duty. By contrast, the existence of a moral duty depends on there being a moral justification for the existence of such rules. Moral duties are not identical to conventional duties, but they are identical to those conventional duties whose existence is morally justified. This is the analysis of moral duties which we gain by substituting a reference to (duty-imposing) conventional rules for Mill's reference to (duty-enforcing) conventional sanctions.

The rationale behind this sort of account is obvious. All we need do is ask why moral duties are not merely conventional duties. The answer to that question is that it is easy to imagine circumstances in which I have the one sort of duty but lack the other. I have a conventional duty but no moral duty (with the same content) if there is a conventional rule imposing the duty on me but it is morally unjustified. The obvious way to exclude this possibility is to require that the rule system which actually applies to me be morally justified. Conversely, I have a moral duty but no conventional duty (with the same content) if a conventional rule imposing the duty on me would be morally justified but there is no such rule. The obvious way to exclude this possibility is to require that morally justified rule systems actually apply to me. We exclude both cases simultaneously if we say that I have a moral duty just in case the imposition on me of a conventional duty (with the same content) is morally justified. Unlike the account of moral duties given by natural duty theories, this analysis includes no reference to a ghostly realm of natural moral rules; the only rules referred to are conventional. None the less, the analysis seems to account easily for the moral force of moral duties; it is provided by the moral justification for the conventional rules imposing the corresponding conventional duties.

Mill's account, suitably amended, thus seems able to explain what it is for a duty to be a moral duty—a duty with moral force. The next step is to extend it to the case of moral rights. Mill provides a simple account of how this extension might go. Recall that his ultimate concern was not with the entire domain of morality (which he regarded as demarcated by the concept of a moral duty) but with a particular sub-domain, namely justice. Mill located this sub-domain by first partitioning duties into two sets: 'duties of perfect obligation are those duties in virtue of which a correlative *right* resides in some person or persons; duties of imperfect obligation are those moral obligations which do not give birth to any right.'[7] He then identified the realm of justice with the realm of perfect obligations, and thus rights: 'It seems to me that this feature in the case—a right in some person, correlative to the moral obligation—constitutes the specific difference between justice, and generosity or beneficence. Justice implies something which it is not only right to do, and wrong not to do, but which some individual person can claim from us as his moral right.'[8]

Mill's argument thus led him from moral duties through moral rights to justice. Since our concern is with rights rather than justice, we need not dwell on the argument's final destination. (Though rights will surely be part of the story in any adequate theory of justice they are unlikely to be the whole story.) But we should note Mill's resultant account of moral rights: 'When we call anything a person's right, we mean that he has a valid claim on society to protect him in the possession of it, either by the force of law, or by that of education and opinion.'[9] There are in this account three distinguishable elements: (1) some form of social protection; (2)

[7] Mill 1969, 247.
[8] Ibid.   [9] Ibid. 250.

having a valid claim to such protection; and (3) some form of sanction which provides the protection. We have already found reason to eliminate the third ingredient in favour of the more general idea of a conventional rule, or a system of such rules. While conventional rules will normally be reinforced by sanctions, such reinforcement is a matter of practical rather than logical necessity. Just as the notion of a conventional duty reinforced by no sanctions is not a logical absurdity, neither is the notion of a conventional right protected by no sanctions. The only form of social protection which is necessarily involved in my having a moral right is my having the corresponding conventional right. Then the second element in Mill's analysis adds the now familiar requirement that the possession of this conventional right be morally justified. If we put all the pieces together the result is a concept of a moral right which is the counterpart of our concept of a moral duty: I have a moral right just in case my possession of the corresponding conventional right is morally justified.

It is worth emphasizing that Mill's account of the nature of moral rights is a piece of conceptual analysis and is therefore entirely independent of his substantive moral theory.[10] Mill himself clearly distinguished between his conceptual and substantive commitments: 'To have a right, then, is, I conceive, to have something which society ought to defend me in the possession of. If the objector goes on to ask why it ought, I can give him no other reason than general utility.'[11] This separation of logically distinct claims is preserved in my reconstruction of Mill's account. I have a moral right, on that reconstruction, just in case my possession of the corresponding conventional right is morally justified. Although this analysis of a moral right is dependent on the notion of a moral justification, it is neutral as regards the shape of that justification. For consequentialists like Mill justifying the existence of a conventional right will consist of two distinct steps. The first step will be to show that either creating or maintaining the right (as the case may be) will promote the theory's basic goals (whatever they might be)—thus Mill's reference to 'general utility'. The second step will be to ground these goals by some method or other (whether realist or constructivist). While the second step is a matter of the theory's methodology—the argument to its basic goods—the first is a matter of its content—the

---

[10] Here I agree with the interpretation of Mill's views in Lyons 1976 and 1977(*a*).
[11] Mill 1969, 250.

argument from these goods to rights. Those contractarians who treat no moral rights as basic will need to supply the analogues of both these steps. But those who treat some rights as basic will justify the existence of the corresponding conventional rights by appealing directly to their methodology—that is, by showing that social recognition of those rights would be one product of the appropriate collective choice procedure. It is because the concept of a moral right is equally compatible with both lines of justification that it is the common property of both consequentialists and contractarians.

Doubtless there are moral frameworks capable of justifying the existence of conventional rights which are neither consequentialist nor contractarian. If so, then this concept will be equally compatible with those frameworks. Indeed, it is even compatible with a natural rights framework. Because such a framework is committed to treating (some) rights as basic, it will have no place for the first step in the consequentialist's justification. But in treating those same rights as objective, it purports to provide its own counterpart to the second step. Whereas contractarians will support their basic rights by appealing to the device of a collective choice procedure, natural rights theorists will rely at this stage on an argument from nature. However, if the critique of this form of argument in the preceding chapter is correct then it cannot succeed, either because the notion of a natural moral rule is incoherent or because no argument from nature can select a determinate set of basic rights. Thus while natural rights theorists can agree that moral rights are morally justified conventional rights, they cannot tell us how the latter are to be justified.

Our interests here are not scholarly; for us Mill has merely pointed the way to what seems a promising resolution of the mystery of how rights can have moral force. However, before we abandon Mill entirely it is worth noting another element in his analysis. From the beginning we have allowed that the concept of a right can admit of competing conceptions. While we have found reason to favour the model of rights as protected choices, Mill inherited from Bentham the competing model of protected interests. It is this model which underlies Mill's view that a duty of justice is one which is owed to an assignable second party (the holder of the correlative right) who will be injured, and thus wronged, in the event of violation of the duty. It thus also underlies his view that 'justice is a name for certain classes of moral rules, which concern the essentials of human well-being more nearly, and are therefore of more absolute obligation, than any other rules for the guidance of life; and the notion which we have found to be of the essence of the idea of justice, that of a right residing in an individual, implies and testifies to this more binding obligation'.[12] The sorts of rules which Mill had in mind were primarily, though not only, those which prohibit, or regulate, physical aggression, the invasion of personal liberty, and the disappointment of legitimate expectations.[13] On the choice conception many of these same rules will count as conferring rights on those whom they protect. But not all will. If the criminal law confers upon me protection against physical assault which I am not free either to waive or to relinquish, then we may say that others have a duty toward me not to commit assault but not that I have a right against them not to be assaulted. The choice conception will thus have the effect of converting some of Mill's rights into mere claims. If we continue to think of the relational duties in these cases as duties of justice, it will also loosen the connection between justice and rights.

The common possession, as between Bentham and Mill, of both a consequentialist moral theory and a conception of rights as protected interests raises the question of why their attitudes toward moral rights were so different.[14] Bentham described the objective of his own work as being 'to enquire, not what are our natural rights, but what in each instance *ought* to be our legal ones upon the principle of utility'.[15] Except for Bentham's preoccupation with the law, this programme is substantively identical to Mill's seeking to ground what he called the rules of justice in 'general utility'. For Mill that was all that was involved in inquiring into our moral rights (or our natural rights, in the purely extensional sense), whereas for Bentham all talk of such rights was anathema. Well, almost all. On rare occasions Bentham did grudgingly concede that talk of moral (or natural) rights might make *some* sense: 'If I say that a man has a natural right to [this] coat or [this] land—all that it can mean, if it mean anything and mean true, is, that I am of

---

[12] Ibid. 255.
[13] Aspects of Mill's theory of justice are explored in Lyons 1978; Gray 1983, chs. 3 and 5; and Berger 1984, chs. 4 and 5.
[14] For one attempt to answer this question see Hart 1982, ch. 4.
[15] Bentham 1952–4, i, 336 n.

opinion he ought to have a political right to it: that by the appropriate services rendered on occasion to him by the appropriate functionaries of government he ought to be protected and secured in the use of it.'[16] The parallels with Mill's account of moral rights are evident. But Bentham in the end resisted the opportunity to take over the language of natural rights: 'I know of no natural rights except what are created by general utility: and even in that sense it were much better the word were never heard of. All such language is at any rate false: all such language is either pernicious, or at the best an improper and fallacious way of indicating what is true.'[17] Thus our question remains.

The answer, I believe, is to be found in the fact that Bentham, unlike Mill, never distinguished between moral rights and natural rights, or between the two senses in which a right might be natural. Since natural rights in the metaphysical sense involve a commitment to a natural rights theory, acceptance of their existence is out of the question for any consequentialist. But rights which are extensionally natural carry with them no such theoretical baggage. If moral rights are compatible with a consequentialist framework—as on Mill's analysis they appear to be—then some of them will almost certainly be extensionally natural. Thus it seems that consequentialists can accommodate not only moral rights but also natural rights. Bentham's rejection of both must again be understood against the backdrop of the political rhetoric of his day. The only appeals to rights with which he was familiar appeared to presuppose a natural rights framework. Since any such framework was unthinkable to Bentham it is understandable that his rejection of these appeals should have led him also to reject the very idea of a natural right (in any sense) and thus also the very idea of a moral right. By Mill's day, on the other hand, the tide of natural rights rhetoric had ebbed somewhat and there was less danger that rehabilitation of the notion of a moral right would give aid and comfort to the enemy. Thus Mill could avow explicitly what (on his analysis) Bentham anyhow allowed implicitly, namely that individuals have such rights.

However this may be, the concept of a moral right which can be extrapolated from Mill's analysis has some obvious virtues, especially for contractarians and consequentialists. Because the

[16] Bentham 1843, iii, 218.
[17] Bentham 1952–4, i, 333.

only ingredients which it requires—conventional rights plus some procedure for justifying such rights—are compatible with both frameworks, it appears to allow contractarians and consequentialists alike to make sense of moral rights and to provide existence conditions for such rights. If this is so, then it enables them to avoid nihilistic conclusions. Since belief in the existence of some rights is deeply embedded in common-sense morality, a nihilistic result for either framework would be highly counter-intuitive. By avoiding such a result this analysis strengthens the case for both frameworks. It does so, furthermore, while side-stepping entirely the questionable metaphysical assumptions of the natural rights tradition. For contractarians and consequentialists alike it appears to be a dream come true: the intoxication of joining the common belief in rights without any nasty hangover. Finally, the analysis also appears to resolve the lingering mystery of how rights can be palpable commodities which we can possess and utilize while also having moral force. They are commodities by virtue of being (actual or possible) conventional rights; they have moral force by virtue of being morally justified.

The analysis does, however, require us to abandon one of our working hypotheses. We began by assuming that all varieties of rights—conventional and moral alike—are so many species of a common genus. We have since complicated this hierarchy by subdividing conventional rights, so that conventional and moral rights now appear to be two families of a common order. Regardless of the precise taxonomical structure involved, the idea was that all of the specific rights in these two categories share a common nature. Our analysis of a moral right has preserved one aspect of this idea. Since moral rights are defined in terms of conventional rights, and since all conventional rights share a common concept of a right as a bundle of Hohfeldian positions, it follows that all rights share this concept. Furthermore, since we have adopted the choice conception of a right we may also say that all rights share this conception. Despite this common ownership of a root notion of a right, however, a biological taxonomy is no longer capable of expressing the relationship between moral and conventional rights. As we can now see clearly, the concept of a moral right analytically presupposes, and is therefore logically parasitic on, that of a conventional right. A moral right therefore stands to its corresponding conventional right not as one member of a natural kind to another but as an operation to the formula which it takes as a value.

The account which we have adapted from Mill tells us that I have a moral right just in case my possession of the corresponding conventional right is morally justified. Although this account is adequate as far as it goes, it needs considerable refinement. We may begin refining it by asking how, for a particular moral right, we are to determine which is the corresponding conventional right. Much of the answer to this question will be given by the characterization of the moral right in question. If that right is determinate, as it should be, then both its scope and its content will be specified. We will thus be told both who holds the right and what it is a right to do (in the case of a liberty-right) or to have done (in the case of a claim-right).[18] The corresponding conventional right must then have the same content and scope. Suppose, for instance, I claim that all human beings have the claim-right not to be enslaved or otherwise treated as mere items of property. Then I am claiming that there is a moral justification for recognizing a right with that same scope and content in some conventional rule system. But which rule system? The answer to this question is not given by my characterization of the moral right.

Since there are many different conventional rule systems, recognition of a particular right, with a particular scope and content, might well be justified within some systems and not others. Most of the rights claims which are prominent in moral/political debate are pleas for the legal recognition of rights and are thus addressed to a municipal legal system. Some may also be addressed to other rule systems. The claim that human beings have the right not to be enslaved is likely to be addressed to all municipal legal systems; it is likely, that is, to include the claim that recognition of this right is justified in all such systems. But it may also include the claim that recognition of the right is justified in the rules of international law, or in those of conventional morality. The existence of some moral rights may thus be a matter of the justifiability of their recognition in a complex network of conventional rule systems. But some rights claims may be addressed only to the municipal legal system. Consider the liberty-right to cast an absentee ballot in a general election. The legal recognition of such a

[18] For simplicity I here ignore the other aspect of the scope of a right, namely its objects.

right might be justified on grounds of fairness or of encouraging a higher electoral turn-out. But it is not a right which needs recognition in any other social institution, nor in a society's conventional morality. Alternatively, some rights claims may not be addressed to the municipal legal system at all. Consider the liberty-right of children (of sufficient age) to have some voice in major family decisions. The recognition of such a right within the rule system of a family might well be justified while its legal recognition would be unjustified (except in certain circumstances, such as custody proceedings).

Rights claims may therefore be (implicitly or explicitly) relative to particular rule systems. If a claim is fully determinate then it will include a specification not only of the scope and content of the moral right in question but also of the rule system (or systems) to which the claim is addressed. In such a case the right claimed is genuine if its recognition in the specified rule system (or systems) is morally justified; otherwise it is spurious. Many rights claims, however, fail to indicate the rule system (or systems) to which they are addressed. In that case the right claimed is genuine just in case its recognition in some conventional rule system is justified; if its recognition can be justified in no such system then it is spurious.

We still need a fuller picture of what it is to justify the recognition of a particular right in a particular rule system. We know that a right is a bundle of Hohfeldian positions whose function is to ensure the right-holder's autonomy over a given domain. I possess a particular right within a conventional rule system only if the rules of that system confer the necessary Hohfeldian advantages on me. The rules of the system do this if they have the appropriate content and are sustained by the appropriate social practices of compliance and acceptance. Our analysis tells us that I have a particular moral right just in case the recognition of that right in some conventional rule system is morally justified. Suppose that it is clear which rule system is in question. Then there are two cases: either that system already recognizes my right or it does not. If it does then what must be justified is the continued recognition of the right, thus the maintenance of the present complex network of social practices. If it does not then what must be justified is altering the system so that it comes to recognize the right, thus the establishment of a new set of practices. We may call each of these options—either maintaining an already existing right or creating a new one—a social policy. A moral right exists when a policy of either sort is justified for the relevant rule system. Demonstrating the existence of a right thus requires that we be able to evaluate social policies.

However, the requirement that a social policy be justified can be interpreted in at least two different ways. On the weak interpretation a policy is justified as long as it is not unjustified or wrong. If we think of any unjustified practice as being morally unacceptable, then a practice is weakly justified as long as it is acceptable. The requirement of weak justification thus establishes a threshold for social policies; as long as a policy surmounts this threshold the distance by which it surmounts it is immaterial. It is therefore possible for each of a number of competing policies to be weakly justified. On the strong interpretation, by contrast, a policy is justified only if it is preferable to all of its rivals, thus if it is not merely acceptable but optimal or ideal. Unless two or more policies are equal best, the requirement of strong justification will select a unique member from a set of competing options. The difference between strong and weak justification is thus the difference between being best and being good enough, between maximizing and satisficing.

It will make a difference which sort of justification we require for the policy of conferring a conventional right. At first glance it seems that a weak justification should suffice. On any plausible account many of the specific moral rights which we possess will be partially dependent on conventional rule systems. This will be true, for instance, of all rights which arise out of agreements. We take for granted that most of the rights which arise in this way have moral force, thus that they are genuine moral rights. However, we would have much less confidence that our various rule systems governing these transactions (the law of contract, for example) are superior to all possible alternative systems. In such cases it seems sufficient that our conventional arrangements not be positively immoral; as long as they pass this (weak) test then the rights to which they give rise have moral force.

Although the option of requiring only a weak justification has its attractions, in the end it must be rejected. Suppose, as seems likely, that there are many alternative conventional arrangements concerning promises and contracts all of which are morally acceptable, but some of which are preferable to others. Each of these alternatives will be weakly justified. If we say that we have a moral right just in case the corresponding conventional right is weakly justified, then each of the alternatives will yield a set of moral rights. However, if they are genuine alternatives then some of the rights which are recognized under them will be mutually exclusive. It will then be true that two or more incompatible moral rights are all genuine. Whatever shape our analysis of moral rights should take, it should not generate inconsistent results.

The problem with weak justification is precisely that it is weak: many incompatible alternative arrangements may all be weakly justified. Existence conditions for moral rights, on the other hand, are intended to supply a test of authenticity for such rights. When confronted with a set of competing rights claims we want to be able to determine which putative right is genuine. We are not aided in this by being told that they all are. The requirement of weak justification is incapable of picking out a unique set of moral rights. We must therefore impose a stronger requirement.

The necessity of requiring a strong justification emerges most clearly for those moral rights whose existence is entirely independent of actual conventional rule systems. These are rights with what we have called a natural criterion, or natural rights in the extensional sense. It seems reasonable to expect a theoretical framework to treat some general rights which are extensionally natural as prior to the more specific rights which depend on conventional rules or roles. After all, one of the traditional functions of moral rights has been to constrain the design of social arrangements, and only natural rights are sufficiently independent of those arrangements to do this job. In the case of natural rights it is particularly clear that inconsistent results are unacceptable. However, if a theory is to yield a unique and consistent set of natural moral rights it must impose a requirement of strong justification. Thus in order to confirm the claim that a particular natural right exists it must be shown that the policy of recognizing it in the appropriate conventional rule system is preferable to any competing policy. Since recognizing a right involves granting the right-holder discretion over the domain defined by the content of the right, competing policies will all consist of alternative ways of regulating that domain. Thus we will here be comparing different possible distributions of freedom or control. The standard to be employed in this comparative evaluation will of course be supplied by a substantive moral theory. Contractarians will therefore ask which of the policies will be selected by some specified collective choice procedure, while consequentialists will ask which of them will best promote some specified goal. It is not the business of a conceptual analysis to settle which of these justificatory mechanisms we should prefer. Whatever our favoured mechanism, what we are seeking is a certain result: namely, that it uniquely select the policy of recognizing the appropriate conventional right.

We will not, however, always get that result. Sometimes the justificatory mechanism, whatever it may be, will be indifferent between two or more alternative policies which it considers equal best. In such a case no policy will be strongly justified, though all the equal best alternatives (and possibly some others as well) will be weakly justified. Suppose that one of these policies confers a right on us which the others withhold. Because this policy is not strongly justified we may not treat this as a genuine moral right. If this seems harsh, we must remember that rights are particularly urgent or peremptory moral considerations. Given their power once they are admitted into moral/political debate, it seems only reasonable to impose admission standards which are not easy to meet. In any case, the equal best alternatives will often converge on some rights while diverging over others. Where this is so, all of the rights in the area of convergence may be considered strongly justified.

The requirement of strong justification applies in the first instance to those moral rights which are also (extensionally) natural rights. These rights, together with other constraints, define the moral positions of individuals independently of any transactions among them. However, since they will necessarily include powers their exercise may create new rights. Thus from a set of general natural rights which constrain social arrangements we may move to specific rights which are the products of those arrangements. It was these rights for which the requirement of weak justification seemed sufficient, since we are inclined to think that the conventional rights which we acquire under our actual social arrangements have moral force as long as those arrangements are not positively immoral. This bias toward the *status quo* is, however, explainable even under the requirement of strong justification, but only once we have further clarified that requirement.

What we know so far is that I have a certain moral right just in case the policy of according that right recognition in some conventional rule system is preferable on moral grounds to all competing policies. But preferable in which circumstances? Suppose I want to know whether I have a moral claim-right to be supplied free of charge with all the ice cream I can consume. There may well be imaginable circumstances of such abundance that a policy of according such a right conventional recognition would be strongly justified. Perhaps under those circumstances I would indeed have the right, but this appears to have little bearing on whether I have it here and now. Justifications of social policies must take into account the actual conditions under which the policies will be implemented. The existence of moral rights is thus not a matter of which rights would be recognized in an ideal rule system under ideal conditions. The policies in question always either sustain already existing rules or create new rules. These options are to be evaluated from the vantage point of the established arrangements; we are always choosing where to go from here. A rights claim provides a moral reason for altering the *status quo* only if it can be shown that the alteration is the best option now available to us. Since the established arrangements will have structured people's expectations, and since reforms may involve disappointing those expectations, there will always be a presumption against reform. (The argument against disappointing expectations may itself appeal to rights.) This presumption is, of course, rebuttable: the gains to be realized from reforming our arrangements will often outweigh the losses. But its existence produces just the bias in favour of the *status quo* which we noted earlier.

The requirement that a conventional right be strongly justified is thus highly contextual. If follows that there is a certain relativity in the concept of a moral right. We have already noted one such relativity: rights claims are (at least often) relativized to particular conventional rule systems. The relativity now in question is to social circumstances. Moral rights which exist under some social conditions may fail to exist under others. For example, some civil or political rights may be justifiable in times of peace and stability but not during a civil upheaval, while some social and economic rights may presuppose favourable material circumstances. Alternatively, certain rights might be more compatible with some cultural traditions than with others. The account developed here is therefore agnostic on the issue of whether any moral rights are universal. Those which respond to relatively fixed aspects of the human condition may indeed belong to everyone alike, while others are socially variable. Whether any given right exists under all actual social circumstances could be determined only by a substantive theory of moral justification.

If we gather together all of the refinements which we have introduced then our analysis of moral rights looks something like this. A moral right with a determinate scope and content is genuine just in case the policy of conferring a right with the same scope and content in some conventional rule system is strongly justified in the actual circumstances under which the system in question would operate. Since this formulation has become rather unwieldy it will not be possible to reiterate it on every occasion. It will therefore be much more convenient to continue saying that a moral right exists whenever the corresponding conventional right is morally justified, or simply that moral rights are morally justified conventional rights. But the full formulation will be what is meant in any such case.

Before we turn to more substantive concerns we should take note of an objection which H. L. A. Hart has raised against Mill's analysis and which applies equally to our own:

> It is plain that an acceptable analysis of moral rights must be such as to allow sense though not necessarily truth to the statement that the existence of such a right is a good moral reason for having and maintaining a law or social convention conferring a right. It must leave room for the assertion for example that the justification for a law conferring a legal right to worship as one pleases is that individuals have a moral right to this freedom. . . . But if . . . to say that men have a moral right to worship as they please already means that there is a good reason why there should be a legal or conventional right to this freedom, the fact that men have this moral right cannot be advanced as a reason why there ought to be such a legal or conventional right. For if the existence of a moral right is to function as such a reason, the fact that such a right exists must be distinct from its being such a reason. If it were not so distinct, the statement that the fact that a moral right exists is a reason why there ought to be a legal right, would when spelt out, amount to the statement that the reason why there ought to be a certain legal right is that there ought to be a legal right.[19]

Hart's diagnosis of the limitation inherent in this sort of analysis is entirely correct. On the account we have developed the claim that we have a moral right to worship freely is logically equivalent to the

[19] Hart 1982, 92.

claim that the policy of conferring the corresponding conventional right on us is morally justified. The former therefore cannot constitute the justification for, or the ground of, the latter. Indeed, the order of discovery is just the reverse: we establish the existence of the moral right by showing that the conventional right is justified, and not the other way round.

This result is, however, not as damaging to our analysis as Hart believes. From a practical point of view what matters most is that we be able to distinguish between those conventional rights which are morally justified and those which are not. While it is reasonable to expect moral rights to play some role in this justificatory process, it matters less which role they play. On our analysis the justification for a conventional right is provided not by the existence of the corresponding moral right but by the resources of a substantive moral theory. It will therefore take the shape of an argument which starts from those resources (whatever they may be) and ends by supporting the policy of conferring the right in question. Suppose that there is some such argument supporting the legal right to worship freely. On our analysis the corresponding moral right cannot appear among the ultimate premises of this argument. (If it did it would simply point to some further justification for the legal right, which would mean that we had not yet reached the ultimate premises.) This is why Hart is correct in saying that on our analysis a moral right cannot constitute the justification for the corresponding conventional right. However, it can, and does, ensure the existence of this justification. The claim that we have the moral right to worship freely is equivalent to the claim that the legal right to do so is morally justified. Its role, therefore, is to assert that there is some valid moral argument which begins with defensible premises and ends by supporting that legal right; it does not, however, provide that argument. On our analysis moral rights are connected to conventional rights not substantively but analytically, not as premises to conclusions but as abstracts to arguments. Rights claims are therefore claims about the relationship between a theoretical framework and a practical policy. They are never self-standing since they merely point to a justification which they do not themselves supply. But the rights which they affirm none the less have moral force, since their existence entails the existence of a moral reason for establishing or maintaining the appropriate practical policy.

Defending a conventional right by appealing to the corresponding moral right is, moreover, neither trivial nor uninformative. A
practical policy can be defended in a number of different ways.
A right, as we have seen, has the function of assuring its holders a
measure of autonomy over some specified domain. The policy of
conferring on individuals the right to worship freely is one possible
way of regulating the domain of religious belief; its rivals will differ
from it precisely in denying individuals this autonomy. If the policy
of conferring the legal right is justified then it must be the best
policy (by some substantive standard) for that domain. Defending
the policy by appealing to the moral right to worship freely
therefore does more than simply assert that it is justified. It also
signals the fact that the salient feature of the policy, in virtue of which
it is justified, is its protection of individual self-determination. A
rights claim thus looks simultaneously back at a justificatory
framework and forward to the function of a conventional right; it
tells us that according to this framework assuring individuals
autonomy over this particular domain is the best available policy. It
therefore points to a particular sort of justification for imposing on
others the constraints which are involved in any right; the value of
these constraints lies in the way in which they safeguard the
autonomy of the individuals whom they protect.

Hart's complaint was that on our analysis an appeal to a moral
right cannot serve as a premiss in an argument justifying the
corresponding conventional right. Thus we cannot make sense of
saying 'Because I have the moral right to worship freely I should
have the legal right to do so.' Hart did not claim that on our
analysis appeals to moral rights cannot serve as premisses in any
moral arguments. Such a claim would be plainly false. Once the
existence of a moral right has been established then that right can
furnish the substantive ground for any number of further conclusions.
If the policy of conferring a legal right to worship freely is justified
then this fact, and therefore the existence of the corresponding
moral right, provides a moral reason for respecting that right on
particular occasions. That reason may well be *prima facie*, and thus
defeasible by competing considerations, but it is not less a reason
for all that. Thus we can make sense of saying 'Because I have the
moral right to worship freely you should not deface my synagogue
with swastikas.' On our analysis, therefore, moral rights have an
important substantive role to play in moral reasoning. The only

limitation of the analysis is that a moral right cannot serve as the ultimate substantive ground for its own conventional recognition. For this ground we must appeal beyond rights.

Perhaps it would be more satisfying to develop an analysis of moral rights which is free even of this limitation. But I cannot see how to do so, and I doubt that it can be done. I strongly suspect that the intuitive force behind the requirement which Hart imposes on an analysis of moral rights reflects the extent to which most of us are still under the spell of the natural rights tradition. If moral rights are conferred on us by a system of natural moral rules then their existence is indeed logically independent of any support which they might provide for conventional rights. The assumption that moral rights are capable of providing the ultimate reasons for conferring the corresponding conventional rights—rather than merely asserting the existence of such reasons—probably results from acceptance of this appealing picture. For all its appeal, however, the picture appears to be incoherent. If it is, and if Hart's requirement depends essentially on it, then we have no choice but to reject that requirement. The analysis developed here may not offer us all we could hope for, but it may offer us all we can have. In any case, it will remain an attractive option until a competitor appears which is capable of satisfying Hart's requirement without resting on dubious metaphysical presuppositions.

5.2 THE CONTRACTARIAN'S DILEMMA

Whereas consequentialists and contractarians agree that moral rights are morally justified conventional rights, they offer rival substantive theories of moral justification. Since consequentialism will occupy us in the next chapter, here we need consider only the contractarian tradition. We have already noted the distinctive contribution of that tradition, namely the thesis that moral principles, including rights principles, are to be justified by deriving them from some hypothetical procedure of collective choice. Although all contractarian theories share allegiance to this thesis, different theories offer different interpretations of the collective choice procedure in question. Thus if we are to assess the adequacy of the contractarian programme for justifying rights, we must attend to issues which divide the theories within the tradition.

A collective choice procedure is a mechanism for transforming the separate wills of some set of agents into a decision which represents their united will. In the contractarian tradition this mechanism has always been understood as an agreement or bargain reached by the agents in question. Theories within the tradition then diverge in their accounts of the nature of this bargaining procedure. For simplicity, let us factor the procedure into two components: an initial situation and a theory of rational choice. The initial situation includes a specification of the nature of the bargaining agents, the circumstances in which they are imagined to be attempting to reach an agreement, the matters to be governed by the agreement, the kinds of information available to the agents, and so on. The theory of rational choice consists of a conception of the rationality of the parties and thus a specification of the normative principles of rational decision-making which they employ in the course of reaching an agreement.

Contractarians have tended to converge on a common theory of rational choice.[20] This theory presupposes that individuals have formed hierarchical structures of ends or preferences. The content of these preferences is settled by each individual for his or her own case; as long as a preference hierarchy satisfies some relatively weak, and largely formal, conditions it is immune to further rational criticism. Rationality is then a purely instrumental matter, exclusively concerned with selecting efficient means for the advancement of one's adopted ends, whatever these may be. Furthermore, it is possible to design a measure for a set of preferences such that the rational choice for an individual in any particular situation can be represented as the option whose selection can be reasonably expected to maximize that measure. If we call this measure utility, then the rational choice is always the choice which maximizes individual (expected) utility, and rational agents are individual utility maximizers.

This interpretation of practical rationality as merely instrumental reflects the contractarian's subjectivism about values. The good or valuable is that which we have adequate reason to desire or pursue. If some goods were objective then their value would be independent of our actually desiring or pursuing them. But in that case their existence would furnish ends which each of us would have adequate reason to adopt regardless of our structure of preferences,

---

[20] See Gauthier 1986, ch. 2, for a representative account.

so that the rejection of these ends would itself be irrational. The instrumental conception of rationality and scepticism about objective values thus go hand in hand.[21] Furthermore, instrumental rationality leads us to a notion of value which is not only subjective but also relative: 'What is good is good ultimately because it is preferred, and it is good from the standpoint of those and only those who prefer it'.[22]

Contractarians tend to regard their reliance on a relatively weak conception of rationality, with its accompanying subjectivism and relativism about values, as a methodological virtue.[23] If our aim is to give morality a firm rational foundation, they argue, then surely we should presuppose nothing more contentious than the minimal conception of rationality capable of doing the job. However, this line of defence misrepresents the options available to us by suggesting that different accounts of practical rationality might lead us to the same moral principles. If this were so then straightforward considerations of theoretical economy would favour the instrumental conception. But it seems much more likely that the theory of rationality which we presuppose as our starting-point will partially determine the moral theory which we derive as our end-point, in which case we will want to take a good look at the latter before selecting the former. The instrumental conception is not our default position simply by virtue of being the weakest conception capable of yielding any moral principles at all; if the only principles it can support are ones we find intuitively repugnant then we will have good reason to try to do better. Of course, if we do decide to seek a stronger conception of rationality we cannot be certain that we will find one for which a plausible independent case can be made. But contractarians have provided no reason for thinking that the task is impossible.

Before we move on, one implication of the contractarian's subjectivism about values needs to be emphasized. On this account what has value for me is determined by what I happen to care about, with no limit imposed on the range of my concerns. While the objects of some of those concerns will doubtless be self-centred, others may well be altruistic or even self-denying. My utility (the satisfaction of all my preferences) must therefore not be confused with my welfare (the satisfaction of my self-centred preferences),

---

[21] See Gauthier 1986, 47 ff.   [22] Ibid. 59.
[23] See, for example, Gauthier 1986, 7–8.

and the maximization of my utility must not be confused with the maximization of my welfare. The instrumental conception of rationality does not identify rational choice with self-interested choice; if it did then it would be stipulating one particular end (the pursuit of their own interest) which individuals are rationally compelled to adopt.[24] This agnosticism concerning the content of individual preferences is sometimes expressed by contractarians as the condition that individuals are mutually disinterested or that they 'take no interest in one another's interests'.[25] But this formulation is seriously misleading. All that it can mean, given the instrumental conception of rationality, is that I behave rationally in so far as I seek to achieve my ends, whatever they may be, rather than yours. But what it also suggests is that my ends include no concern for your pursuit of yours. On the former interpretation the condition merely stipulates the ownership of the set of preferences whose satisfaction I am assumed to be seeking (they must all be mine); on the latter it additionally stipulates their content (they must all be about me). But stipulations about the content of individual preferences are precisely what the instrumental conception disallows. Just as contractarians wish not to presume that individuals are sympathetic or benevolent, they must also not presume that they are exclusively self-interested.[26]

In the contractarian scheme, therefore, individuals who bring to the initial situation diverse sets of aims and ambitions are able to locate some common set of moral principles which it is rational for each to accept. This procedure of collective choice is then advanced as a justification for those principles; we are to acknowledge them as principles binding on us because they would be agreed on by those agents in that situation. This is the sense in which, as contractarians are wont to say, a theory of morality is a part of the theory of rational choice.[27] But this methodology raises an obvious question: why should I acknowledge as morally justified the set of principles which would be accepted by those agents in that

---

[24] In terms of the categories developed in Parfit 1984, contractarians are committed to defending some version of the Present-aim Theory of rationality against the Self-interest Theory. Cf. Gauthier 1986, 32 ff.

[25] Rawls 1971, 147.

[26] As is recognized by Gauthier when he speaks of 'persons who may take no interest in others' interests' (Gauthier 1986, 17); cf. Narveson 1984, 170–1.

[27] See, for example, Rawls 1971, 16; Gauthier 1986, 2–3. But for a second thought see Rawls 1985, 237 n.

situation? Part of the answer to this question is provided by the theory of rational choice: these are principles which those agents do not merely happen to accept but which they are rationally required to accept. Let us assume, for simplicity, that we know which principles of choice rational individuals will employ under the conditions of the initial situation, and thus also which moral principles they will come to agree on. It is still unclear why that result should matter to me—why I should regard myself as morally bound by the set of principles on which they would converge. After all, the collective choice procedure which yields these principles as output is entirely hypothetical. It may be true that I am morally bound by the agreements I have actually made, but it is doubtful that I am bound by any agreements I have not actually made, even if it would have been rational in some circumstances or other for me to make them. And I am certainly not bound by any agreements which would have been rational for some set of imagined agents in some imagined circumstances. As Ronald Dworkin has justly observed, 'a hypothetical contract is not simply a pale form of an actual contract; it is no contract at all'.[28]

The problem with the contractarian's reliance on a hypothetical collective choice procedure is that it is hard to see why the fact that certain conventional rights would be selected by such a procedure should count as a moral justification of them. Since it is clear that the theory of rational choice cannot provide the full solution to this problem, we must have recourse to the other component of the choice procedure: the initial situation. The basic idea behind a contractarian methodology is that we should acknowledge as morally justified whatever social arrangements would be agreed on in a suitably characterized initial situation. The methodology is thus purely procedural; there are no independent moral standards which the product of the agreement must satisfy.[29] This methodology obviously rests a good deal of moral weight on the interpretation of the initial choice situation. However, whereas contractarians have tended to converge on a common theory of rationality, they have offered very different interpretations of this situation.

Recall that the initial situation stipulates the various conditions

---

[28] Dworkin 1977, 151. Some contractarians are aware of this problem with their methodology; see, for instance, Narveson 1984, sect. 4.
[29] See the account of pure procedural justice in Rawls 1971, 85–7, 120; cf. Buchanan 1975, 164, 167.

under which rational agents are imagined as attempting to reach an agreement: what their world is like, how they have interacted with one another, what kinds of information are available to them, and so on. All contractarian interpretations of the initial situation share some features, such as the assumption that the circumstances of justice obtain.[30] Beyond this common ground there is room for a good deal of variety. For our purposes, however, we need examine only one dimension of disagreement. Let us call a construal of the initial situation moralized if it imposes any moral constraints on the interactions of the parties and non-moralized if it does not. Every contractarian theory must presuppose either a moralized or a non-moralized initial situation, and there are well-known examples in the tradition on each side of this divide. However, neither option appears to be capable of explaining why the contractarian's procedure of hypothetical collective choice should be regarded as a self-standing method of moral justification.

Consider first a non-moralized initial situation. The classic case is that of Hobbes, who regarded justice (by which he meant all moral duties and rights) as being first established by the agreement which rational individuals would reach in order to extricate themselves from the state of nature and its war of all against all. From this he drew the obvious conclusion:

> To this warre of every man against every man, this also is consequent; that nothing can be Unjust. The notions of Right and Wrong, Justice and Injustice have there no place.Where there is no common Power there is no Law: where no Law, no Injustice. Force, and Fraud, are in warre the two Cardinall vertues.[31]

We are therefore to imagine individuals whose state of nature interactions have been governed by no moral constraints, meeting to seek agreement on a set of principles which are to regulate their common affairs. Each bargains with the others in full knowledge of the position which he/she has been able to achieve and maintain through the use of force and fraud. The outcome of this agreement is then to be acknowledged, by us as by them, as a system of moral rights and duties.

A more recent theory of the Hobbesian variety may be found in the work of James Buchanan.[32] In Buchanan's scheme the device of

---

[30] A standard account may be found in Rawls 1971, 126–30.
[31] Hobbes 1968, 83.   [32] Buchanan 1975.

a hypothetical agreement is used to establish a set of basic moral rights which can then be utilized by their holders to generate further rights. The situation in which this agreement is imagined as being reached is what Buchanan calls the 'natural distribution': the distribution of assets which has resulted from prior unregulated interaction in the state of nature. This distribution is unlikely to be egalitarian: 'There is nothing to suggest that men must enter the initial negotiating process as equals. Men enter as they are in some natural state, and this may embody significant differences.'[33] Just how significant the differences might be is revealed by Buchanan's observation that in a condition of anarchy 'some persons may have the capacities to eliminate others of the species. In this instance, the natural equilibrium may be reached only when the survivors exercise exclusive environmental domain.'[34] However this may be, the inequalities inherent in the natural distribution will be transmitted to the emergent structure of moral rights:

The specific distribution of rights that comes in the initial leap from anarchy is directly linked to the relative commands over goods and the relative freedom of behavior enjoyed by the separate persons in the previously existing natural state. This is a necessary consequence of contractual agreement. In Hobbes's model, there are, by inference, considerable differences among separate persons in a precontract setting. To the extent that such differences exist, postcontract inequality in property and in human rights must be predicted.[35]

It seems scarcely worth pointing out that theories constructed on this Hobbesian model are incapable of solving the root problem inherent in a contractarian methodology. We want to know why we should accept the conventional arrangements which would emerge from a hypothetical agreement as morally justified, thus why we should regard the conventional rights which such an agreement would establish as having moral force. We are offered no answer whatever to these questions if we are told that these arrangements and these rights would be agreed to by a set of agents who have been antecedently free to engage in unbridled force and fraud. In the war of all against all the strong and ruthless will tend to prevail. One of the functions of a set of moral rights is to negate the advantages of predation by providing equal protection for all. But equal protection will not be the outcome of a rational bargain if

[33] Ibid. 26.   [34] Ibid. 59.   [35] Ibid. 25.

the strong and ruthless are allowed to bring their ill-gotten gains to the bargaining table. The Hobbesian model may well be apt as an analysis of *realpolitik*, but why we should regard it as a method of moral justification remains utterly mysterious.

If we are to acknowledge the conventional rights which would emerge from a hypothetical agreement as having moral force then some controls must be imposed on the conditions under which the agreement is imagined as being made. We are brought, therefore, to the option of a moralized initial situation. The most celebrated example of a contractarian theory which has taken up this option is to be found in the work of John Rawls. The principles of justice which Rawls seeks to justify can be regarded as distributing basic moral rights (perhaps among other advantages).[36] We need not concern ourselves however with their content, but only with their derivation from the rational choice of a set of individuals in an initial situation. Rawls is aware of the fact that different interpretations of this situation will yield different versions of a contract theory. His own interpretation—what he calls the original position—is deliberately constructed so as to embody standards of fairness:

By contrast with social theory, the aim is to characterize this situation so that the principles that would be chosen, whatever they turn out to be, are acceptable from a moral point of view. The original position is defined in such a way that it is a status quo in which any agreements reached are fair. It is a state of affairs in which the parties are equally represented as moral persons and the outcome is not conditioned by arbitrary contingencies or the relative balance of social forces. Thus justice as fairness is able to use the idea of pure procedural justice from the beginning.[37]

It is now obvious why Rawls calls his theory justice as fairness: the principles of justice (including rights) are those which would be chosen in a fair initial situation. Rawls construes a fair situation as one in which the bargaining agents are free and equal. The principal device for ensuring their freedom and equality is the veil of ignorance: by denying the agents knowledge of their natural assets and social positions Rawls deprives them of the opportunity to exploit these to their advantage. By effectively removing all of the contingencies which differentiate individuals, Rawls also removes

---

[36] For an exposition of Rawls's theory of justice as a theory of rights see Martin 1985.

[37] Rawls 1971, 120.

the basis for genuine bargaining among them, thereby reducing a problem in game theory to one in individual decision-making under uncertainty.

Assume that individuals in Rawls' original position would indeed choose his principles of justice. If we ask why we should acknowledge those principles as morally binding, Rawls has an answer which was unavailable to Hobbes and Buchanan: the principles which rational individuals would agree on under fair conditions are themselves fair. But this answer merely raises another question for the contractarian: how are these antecedent standards of fairness to be justified? Since they are presupposed in the theory's collective choice procedure they presumably cannot be a product of that procedure. But then they must be given some other justification, which cannot itself be contractarian.

It is important to note that this result does not depend on the substantive standard of fairness which Rawls employs to construct his original position. Although the case which Rawls makes for his conditions of freedom and equality has been much disputed, we need not call it into question.[38] Nor need we worry about its deep structure. Ronald Dworkin has argued that Rawls's conception of procedural fairness presupposes the basic right of individuals to equal concern and respect, so that his theory of justice is really right-based.[39] If this diagnosis is correct then Rawls's theoretical framework contains some rights which are not themselves the products of any collective choice procedure, and so it fails to be a contractarian rights theory.[40] But even if Dworkin is mistaken, the fact remains that some substantive components of Rawls's framework are not contractually based. Rawls defends the moral constraints which he imposes on the original position as reasonable and argues that they can be brought into reflective equilibrium with the conception of justice to which they lead. However this may be, it is at odds with the contractarian claim that standards of justice and fairness are posterior to, because derived from, a collective choice procedure. The contract mechanism in Rawls's theory appears to have the status of a mere expository device.

If we gather together our results to this point then the dilemma

---

[38] For one form of critique (from the right) see Nozick 1974, 183–231, and Gauthier 1986, 245–54.

[39] Dworkin 1977, ch. 6.

[40] The diagnosis is rejected by the patient in Rawls 1985, 236 n.

confronting the contractarian rights theorist is obvious. For the contractarian conventional rights are morally justified only if they can be derived from an appropriate collective choice procedure. This procedure must specify the initial situation in which agreement on a set of social arrangements, including a set of conventional rights, is imagined as being reached. An interpretation of this initial situation either includes moral constraints or it does not. If it does not then the rights agreed on in the situation will have no moral force. But if it does then those constraints themselves will have no contractarian foundation. As a moral foundation for rights the contractarian methodology is therefore either irrelevant or incomplete. In either case the programme of justifying moral rights by deriving them exclusively from an agreement among rational individuals appears to be doomed to failure.

This may not, however, be the end of the story. An interesting attempt to evade this dilemma can be found in the work of David Gauthier.[41] Gauthier accepts the contractarian programme of deriving a moral theory from the theory of rational choice. Furthermore, he insists that the premisses of this theory of rational choice must not themselves contain any moral ingredients.[42] In this he appears to side with Hobbes and Buchanan against Rawls. It is somewhat surprising therefore to discover that his characterization of the initial situation includes a moral condition, which he calls the Lockean proviso. The proviso regulates interaction among individuals in the state of nature by requiring them not to better their own condition by worsening the condition of any other. It has the force of a moral principle because it is capable of constraining individual utility maximization. Furthermore, its theoretical function is to guarantee that the initial bargaining situation is fair or impartial. Unlike Buchanan, Gauthier recognizes that 'fair procedures yield an impartial outcome only from an impartial initial position'.[43] The proviso is thus meant to ensure that the social arrangements agreed on in the initial situation will have moral force.

The effect of the proviso is to assign individuals in the imagined state of natural interaction a set of property rights which begin in their own persons and end by extending into the external world.

---

[41] Gauthier's views were developed in a lengthy series of articles, but these have all been superseded by Gauthier 1986.

[42] 'Morality, we shall argue, can be generated as a rational constraint from the non-moral premisses of rational choice' (ibid. 4).　　　　　[43] Ibid. 191.

Gauthier's construal of the initial situation thus appears to be closer to Locke's state of nature than to Hobbes's. Like the former and unlike the latter, Gauthier holds that some moral rights antedate all agreements: 'They are what each person brings to the bargaining table, not what she takes from it.'[44] This claim seems to commit him to accepting some moral principles which are not themselves contractually based, thus siding with Rawls against Hobbes and Buchanan. Moreover, it seems obviously inconsistent with his own requirement that morality be entirely derivable from the non-moral premisses of the theory of rational choice. What exactly is happening here?

The answer lies in Gauthier's claim that it is rational—that is, utility-maximizing—for each individual to accept the proviso as a constraint on natural interaction, in advance of any agreement. If this claim can be justified, then all of the ingredients in Gauthier's moral theory will be derivable from the theory of rational choice, even though they will not all be derivable from a collective choice procedure. Such a theory would surely still deserve to be called contractarian despite its reliance on a moralized initial situation. The rational support which Gauthier claims for the proviso thus promises to steer him safely between the horns of the contractarian's dilemma.

But can the proviso be supported in this way? Will compliance with it be the utility-maximizing strategy for each agent under conditions of natural interaction? Actually, the claim which Gauthier defends is weaker than this in two respects. First of all, it is not unconditional compliance which is rational for each agent but only compliance when enough others are complying as well.[45] Thus each person's maximizing strategy is a conditional one, and general compliance with the proviso has the status of a convention which will be relatively stable because it promises to facilitate a further co-operative agreement which will be beneficial to all. However, even in this more guarded form the claim invites an obvious rejoinder. While conditional compliance may be the maximizing strategy for many, unconditional violation may be the maximizing strategy for some. But if this is so then there will be some agents in the initial situation with no reason not to exploit others.

---

[44] Ibid. 222. [45] Ibid. 193.

Gauthier does not deny that this will be so; this is the second respect in which his claim has been weakened.[46] Nor would any such denial be plausible. When we bear in mind the very considerable disparities among individuals, both in their aims and in their natural abilities, it is difficult to believe that any single strategy—whether conditional compliance with the Lockean proviso or any other—will be uniquely maximizing for all. Gauthier's invocation of the proviso invites us to contemplate a peaceable kingdom in which because no one has any reason to exploit others everyone is safe from being exploited. But if individuals are guided only by instrumental rationality—that is, by a calculation of their own aims and abilities—then we must expect some division between the exploiters and the exploited. Gauthier may be right in thinking that rational individuals in the state of nature will manage something better than Hobbes's war of all against all. But as long as instrumental rationality permits some exploitation by the strong of the weak then natural interaction will fail to be a fair initial situation for agreement on a set of rights. The requirements of instrumental rationality thus appear to be too weak to yield a set of rights with moral force. The rights which are generated by the theory of rational choice are ones we have no reason to acknowledge as morally justified.

Earlier I said that the instrumental conception of rationality, and the subjectivism about values which accompanies it, should be regarded as working assumptions in a moral theory which will be confirmed to the extent that the theory goes well and disconfirmed to the extent that it goes badly. Contractarian theories of rights have not gone well. If they do not ensure the fairness of the initial choice situation then they will have to recognize that any rights agreed on in that situation will lack a moral justification, while if they do ensure it then they will have to acknowledge that at least some values are objective. While the former option is cynical the latter is suicidal. In either case we have ample motivation to explore the possibility of grounding rights in a more ambitious conception of rationality and a set of avowedly objective values.

[46] Or so I interpret Gauthier's brief remarks on technological dominance (ibid. 231).

6

Consequentialist Rights

IT is time to take stock of results so far and to remind ourselves of the task at hand. The spiralling inflation of rights rhetoric led us to seek a set of existence conditions which would include both an analysis of moral rights and a substantive procedure for verifying claims about such rights. We now have the first of these resources in hand: a moral right is a morally justified conventional right. This analysis has been constructed out of three ingredients: the model of rights as protected choices, a set of existence conditions for conventional rights, and the notion of a moral justification for such rights. Because a concept of a moral right should be compatible with different normative theories of rights, the third of these ingredients does not specify how conventional rights are to be morally justified. The project of developing a verification procedure for rights claims is thus reduced to that of providing an adequate substantive account of moral justification. This account has so far eluded us. Our critique of natural rights theories yielded the conclusion that moral rights cannot be both objective and basic. It therefore left two possibilities open: either such rights are subjective or they are derivative (or both). Pursuing the first option led us to contractarianism. The problem with the contractarian account, however, is that it is difficult to see why the conventional rights which are derivable from a hypothetical agreement made in a Hobbesian state of nature should be regarded as having moral force. We therefore seem no closer to the substantive theory of justification which we need in order to complete our existence conditions for moral rights.

Since there may well be versions of a subjective theory of rights which do not rely on the device of a hypothetical contract, we have not succeeded in closing off the first option entirely. However, contractarianism has been by far the best developed and most influential subjective theory; its failure therefore gives us ample reason to switch over to the second option. Here too we will limit ourselves to considering the best developed and most influential alternative, namely consequentialism. Since there may well be other theories which deny that rights are basic, our inquiry in this chapter will once again fall short of being exhaustive. However, it will complete our examination of the three most prominent theoretical frameworks in contemporary moral philosophy. Of these three only consequentialism remains as a way of avoiding nihilistic conclusions about rights.

Since consequentialist theories have a reputation for hostility to rights, nihilism may now seem unavoidable. The main burden of the present chapter is to show that this despairing conclusion is premature. Our task will be complicated by the fact that, like their contractarian counterparts, consequentialist theories admit of much variety. However, the results of the preceding chapter will enable us to reduce this variety in one important respect. As we have seen, contractarian theories are identified by their method: they leave open the question whether rights are basic while denying that they are objective. Consequentialist theories, by contrast, are identified by their content: they leave open the question whether rights are objective while denying that they are basic. Because of this division between method and content, we found it convenient in our examination of contractarianism to confine our attention to those versions which treat rights as basic. It will be similarly convenient in our examination of consequentialism to confine our attention to those versions which treat rights as objective. Indeed, it will be more than convenient. If contractarian theories are incapable of providing a moral justification for conventional rights, then those versions of consequentialism which employ a contractarian methodology will fare no better.[1] It appears that the only viable consequentialist theories are objective theories.

This still leaves a formidable array of theories for us to contend with. Since we are interested in the idea of a consequentialist moral structure as a framework for rights, we do not wish to confine our attention to any specific theory which displays this structure, such as utilitarianism. For our purposes the features which differentiate the various species of the genus are quite irrelevant. However, we do need an account of the nature of the genus; our first task, therefore, must be to construct such an account.

[1] Thus I exclude the contractarian route to utilitarianism followed in Harsanyi 1982 and 1985.

6.1 CONSEQUENTIALIST GOALS

What makes a theory a consequentialist theory? It seems likely that the character of a moral theory is determined by the character of its basic principles. What, then, is distinctive about the basic principles of a consequentialist theory? If we take the label itself literally we will be led to something like the following account. Suppose, for simplicity, that the basic principles of a moral theory appraise actions as good or bad, right or wrong. Then a theory is consequentialist just in case it appraises actions solely in terms of the value of their consequences. It follows that non-consequentialist theories appraise actions at least in part in terms of some aspect of them other than their consequences, such as their intrinsic nature.

This textbook account seems intuitively to point in the right direction. After all, we expect non-consequentialists to hold that some actions are intrinsically wrong—wrong by their very nature—whereas consequentialists are committed to thinking that such actions would be justified in circumstances in which their outcomes would be for the best. The account also captures our sense that consequentialist theories are basically theories of the good and only derivatively theories of the right. We expect consequentialists to begin by giving us an inventory of ultimate or intrinsic values: those states of affairs which are valuable for their own sake and not for the sake of any further states which they produce or promote. This inventory can then be used to evaluate the outcomes of our actions, and the rightness or wrongness of an action can be made a function of the value of its outcome. Thus no action can be intrinsically wrong, if this means that it would be wrong whatever the value of its outcome.

Despite its intuitive appeal, however, this account is surprisingly resistant to further development. For one thing, if it is to apply to all forms of consequentialism then it will have to broaden the range of appraisals which can be made by basic moral principles. If consequentialists appraise actions solely in terms of their consequences then they do the same for motives, intentions, dispositions, decisions, policies, strategies, lives, institutions, societies—in short, for all those manifestations of agency which are fit subjects for moral evaluation. Furthermore, in each of these cases the account will have to loosen the connection between the moral value of a thing and the value of its consequences. For instance, whereas some consequentialists think that the rightness or wrongness of an action is determined by its own consequences, others hold that it is determined by the consequences of the general practice of actions of that kind, or the consequences of general compliance with a rule requiring actions of that kind, or the consequences of instituting and enforcing such a rule, or whatever. If these latter views are to count as consequentialist then we will need to recast our initial account into something like the following form: a theory is consequentialist just in case its basic principles appraise every aspect of moral agency solely in terms of the value of some set of consequences.

But these problems with the account are relatively minor. The real difficulties surface when we try to specify what the account excludes as grounds of appraisal, thus how a theory might manage not to be consequentialist. Our initial formulation suggested a boundary between the intrinsic nature of an action and its consequences. But this boundary is notoriously difficult to locate. Specifications of what an agent does on a particular occasion are sufficiently elastic that what will count on one as a part of the intrinsic nature of the action will count on another as one of its consequences. Thus what I do may be described variously as making marks on a piece of paper, signing a cheque, paying a bribe, or ensuring the survival of my business. The more expansive of these alternative characterizations incorporate into the intrinsic nature of my action states of affairs which the more austere expel into its consequences. Establishing a firm boundary between acts and their outcomes would thus seem to require showing that there is some favoured or privileged specification for every act. It is unclear how, or whether, this can be done. To make matters still worse, on some conceptions of a consequence the boundary between acts and their outcomes vanishes entirely. Consequentialists find attractive (or at any rate should find attractive) the idea that the outcome of an action consists of the net difference it makes to the world, in which case it includes all and only those states of affairs which occur if it is done and do not occur if something else is done instead.[2] However, one state of affairs which will occur if and only if I do something is the fact that I have done that thing. On this conception of a consequence, therefore, every act becomes, trivially, a consequence of itself.

[2] See Parfit 1984, 67–70.

Perhaps consequentialists and their opponents really do embrace different theories of action and causation, so that if we were to pursue the foregoing issues we would be led in the end to the deep difference between their moral outlooks.[3] But I think we will do better to abandon this line of inquiry altogether and start over. Suppose we focus instead on the fact, which we have already noticed, that the basic principles of a consequentialist framework are principles of the good rather than of the right. This structural feature is reflected in the familiar observation that consequentialists affirm the priority of the good over the right.[4] This priority has two dimensions: (1) a theory's principles of the good presuppose no antecedent principles of the right; and (2) the right is then construed as that which promotes or produces the good. Satisfaction of both of these conditions is guaranteed if a theory is good-based—that is, if its basic principles are all principles of the good. Because these principles are basic, they cannot rest on any deeper principles of duties or rights. And because no other principles are basic, if the theory contains any principles of duties or rights then these must be derived from its principles of the good. Perhaps, then, it is the fact that it exemplifies this moral structure which makes a theory a consequentialist theory.

Acceptance of the priority of the good is certainly a necessary condition for a theory to count as consequentialist, but it is not sufficient. Every consequentialist theory begins with a theory of the good, but so do some non-consequentialist theories. Consider, for example, the moral framework defended by John Finnis.[5] Finnis himself calls this framework a natural rights theory, but it fails to satisfy our earlier characterization of such theories by virtue of treating goods rather than rights as basic. However, although he does go on to build a theory of the right on his theory of the good, Finnis would quite justifiably reject the suggestion that his framework is therefore consequentialist. If we are to isolate consequentialist theories we must therefore draw a boundary somewhere within the class of theories all of which agree that the good is prior to the right.

What we have so far failed to capture is the fact that consequentialists are committed to the pursuit of some synoptic or global goal, thus the fact that their moral framework is not merely

---

[3] Some think so; see Donagan 1977, 37–52.
[4] Rawls 1971, 24–32.   [5] In Finnis 1980 and 1983.

good-based but also goal-based. It will be best to construct an account of consequentialist goals in three distinct stages. The first stage is the one which we already have on the agenda: a basic theory of the good. We have described such a theory as an inventory of those states of affairs which are valuable in themselves, thus worth pursuing for their own sake. We might, alternatively, characterize these goods as the states which we have adequate or sufficient reason to pursue, or at least to aim at or desire, for their own sake. When put this way, the theory of value which is the consequentialist's starting-point seems to resemble that of the contractarian. The resemblance is, however, misleading. As we noted in the last chapter, the contractarian's conception of value is both subjective and relative. It is subjective because the value of any state of affairs is dependent on its being valued or desired or aimed at by some agent. And it is relative because such states are valuable only for those who have the appropriate attitude toward them. By contrast, the consequentialist's conception of ultimate or intrinsic value is neither subjective nor relative.[6] It is not subjective because the favoured states of affairs are valuable whether or not they are valued or desired or aimed at by some agent, or indeed by any agent. And it is not relative because these states are not merely valuable for some particular agent: they are valuable, period (or, alternatively, valuable for any agent).

If we assume a connection between the valuable and the rational, then these contrasting conceptions of value require contrasting conceptions of rationality. We have already explored the contractarian's coupling of subjectivism about values and a thin theory of instrumental rationality. If consequentialists wish to affirm that the good is what we have adequate reason to aim at, or that it is what will be pursued by any rational agent, then they will need a conception of rationality which enables them to appraise ends as well as means. The substance of a consequentialist theory of the good is that there are certain states of the world which everyone ought to pursue, and that either indifference or hostility to these states is contrary to reason. Thus whereas contractarians hold that states of the world are valuable only if some agent has reason to pursue them, consequentialists hold that there are some states which all agents have reason to pursue because they are valuable.

---

[6] Unless, of course, the consequentialist is also a contractarian.

For contractarians the rational entirely determines the valuable, while for consequentialists the valuable partially conditions the rational (or reasonable). Contractarians are, of course, sceptical concerning the prospects of developing both this objective conception of value and the thicker theory of practical rationality which is its natural companion. But since we are interested for the moment only in understanding the nature of consequentialism, the question whether this scepticism is well- or ill-grounded need not detain us.

We should, however, pause to note the implications of embracing a non-relative conception of value. For contractarians facts about ultimate goods have the following form: this state of affairs is valuable for, or to, or from the point of view of, this agent. It follows that some state may be valuable from my point of view but not from yours, thus that I may have a reason to pursue or promote it while you have none. This relativity of the value of a state to a valuing agent, it should be stressed, is independent of any further relationship between the state and the agent. In particular, a state which is valuable *for me* need not also be a state *of me*, or indeed of anyone (this is guaranteed by the absence of content restrictions on preferences). Whatever (or whomever) valuable states may be states of, their value is always relative to the concerns of some agent or group of agents. The constrasting conception of value embraced by consequentialists omits this relativity. For consequentialists facts about ultimate goods must have the following form: this state of affairs is valuable. Or alternatively: this state of affairs is valuable for, or to, or from the point of view of, anyone. It follows from this conception that if a state is valuable from anyone's point of view then it is valuable from everyone's point of view, thus that if anyone has a reason to pursue or promote it everyone does. Again, this feature of the value of ultimate goods is independent of their location or ownership. Suppose, for example, that health is an ultimate value. If value is relative then only those who happen to care about my health will have any reason to promote it, whereas if value is non-relative then everyone will have such a reason, whatever the content of their concerns. A non-relative conception of value thus enables us to make sense of saying that there are certain ultimate goods—life, health, knowledge, pleasure, or whatever—which all of us have reason to promote, wherever they may be found and whomever they may belong to.

This feature of ultimate goods is part of what is meant by saying that they are agent-neutral (by contrast with the agent-relativity of the contractarian's values).[7] This agent-neutrality captures something of the impartiality which we expect of a moral framework: we are to take seriously not just our own goods or those of individuals about whom we happen to care, but everyone's. And it also partially explains the sense in which the demands of morality are categorical: the promotion of these goods has a rational claim on us whatever our personal ends or preferences may happen to be. In any case, all theories in the consequentialist tradition begin by affirming the existence of some set of ultimate goods which are agent-neutral in this sense. They are then differentiated in part by their divergent theories of value. There is plainly room for much variety here. Some theories may be monistic, affirming the existence of but one ultimate good, while others are pluralistic. For some theories ultimate goods may be limited to states of experience, while others cast the net more widely. Some theories may confine themselves to states of individuals, while others allow that collectivities can have their own irreducible goods. And so on and on. Luckily, our concerns do not require us either to sort these various options or to arbitrate among them. Our immediate aim is merely to distinguish consequentialist theories as a group from their rivals, and our ultimate aim is to determine whether the structure of such theories enables them to provide a moral foundation for rights. These aims will be well enough served by noting that a consequentialist framework must begin by treating some ultimate goods as objective and agent-neutral; we need not worry about which goods are to play this foundational role.

A theory of value provides the building blocks for a consequentialist goal. The second stage in constructing such a goal consists of specifying some operation for combining the theory's separate basic goods into a single global value. This operation will be necessary whether a theory of value is monistic or pluralistic. If it is monistic then the goods to be collated will be the several instances of the theory's one ultimate generic value. Thus, for example, in a utilitarian framework which treats nothing but individual welfare as ultimately valuable these goods will consist of distinct costs and benefits (either of the same individual or of different individuals). If a theory is pluralistic, on the other hand, then it must additionally

[7] See Nagel 1986, 152–3, 158–63.

provide some means of collating its different generic values. Thus, for example, a perfectionist framework which treats a number of excellences as ultimately valuable must supply some operation for combining the several instances of its several categories of goods.

Any operation for combining separate goods into a single global value will presuppose the commensurability of these goods. Assume, for simplicity, that the only ultimately valuable states are states of individuals. A monistic theory of value will require both intrapersonal and interpersonal comparisons of specific instances of its sole good, while a pluralistic theory will additionally require intercategorial comparisons among its several goods. We need not be too demanding here: a consequentialist framework can probably tolerate some incommensurable values. But it will certainly require at the very least that most of its basic values be commensurable in most respects most of the time. This will be so regardless of the particular combinatory operation which is chosen to define a consequentialist goal. Once again consequentialists have a number of options available. Because it has been favoured by utilitarians, the most familiar operation in the consequentialist tradition is arithmetic addition or aggregation. But consequentialists can also choose operations which are wholly or partially distributive: an equal distribution of goods across individuals, or a pattern which attends solely to the minimum individual share, or an operation which is sensitive to both aggregative and distributive factors, or whatever. The range of operations which could in principle be used to generate a single global value from a set of particular inputs is bounded only by the limits of our mathematical ingenuity. Doubtless many of these possible options would generate rather bizarre moral theories, but they would all be consequentialist theories.

The third stage in the construction of a consequentialist goal merely completes the second. If a goal is to furnish practical guidance it is not enough simply to define a global value: we must also be told what to do about or with that value. Since it is formed from a set of particular goods it is presumably to be promoted rather than retarded. But to what extent are we to promote it? The answer to this question will consist of some function defined on the global value. Again because it has been favoured by utilitarians, the most familiar such function in the consequentialist tradition is maximization. If we choose this option we will always prefer more of the global value to less and in any context of choice we will aim at bringing about as much of it as possible. A maximizing function thus leads us to a moral ordering of available alternatives in terms of the extent to which they will promote the global value in question, the best alternative being the maximizing one. Maximization is such a standard function for consequentialist theories that it is easy to forget that it is not the only available option. Once again the range of possible functions has mathematical rather than moral limits. But some non-maximizing functions, such as satisficing, might result in an intuitively attractive consequentialist goal.[8]

If we draw together the three stages we have outlined then we may say that a consequentialist goal consists of (1) some set of basic, ultimate, objective, agent-neutral goods; (2) some operation for combining these separate goods into a single global value; and (3) some function which specifies how this value is to be promoted. We have already noted that the first stage is not unique to consequentialists. But the next two stages jointly define the global or impersonal point of view which is the characteristic signature of the consequentialist tradition. Although a consequentialist goal begins with a set of particular values each of which has a different location in the world, it ends by merging them into a single standpoint which transcends all of these separate locations. It is this transcendent perspective—what Thomas Nagel has called 'the view of the world from nowhere within it'—which determines all basic moral appraisals in a consequentialist framework. There is thus a further respect in which a consequentialist goal is agent-neutral. Recall that the particular goods from which that goal is constituted are each agent-neutral by virtue of providing reasons for action for any agent whatever. The global value into which they are processed is also agent-neutral by virtue of providing a common ultimate aim for all moral agents.[9]

Resistance to consequentialism typically takes the form of rejecting the hegemony of this impersonal standpoint. Just as Finnis's positive views showed how consequentialist and non-consequentialist theories can both be good-based, his critique of consequentialism shows where the two frameworks must part company. Having identified a set of basic goods, Finnis then argues that combining them into a single determining global value would

[8] For a defence of a satisficing version of consequentialism see Slote 1985, ch. 3.
[9] See Parfit 1984, 27.

be both irrational and immoral.[10] Having joined the first stage in the construction of a consequentialist goal, Finnis declines to continue through the next two. Instead, he proposes a theory of the right on which it is impermissible to damage any particular instance of any of his basic goods. In this framework actions are appraised for their impact on each instance of each good considered separately; there is no global standpoint capable of overriding or adjusting these several particular standpoints.

It was the dominance of this transcendent point of view toward which we were groping in our initial characterization of consequentialist theories. Distinguishing consequentialists from their rivals does not require distinguishing actions from their consequences. Consequentialists need not care what we count as belonging to the intrinsic nature of an action and what we count as belonging to its outcome. Their claim is that among all the aspects of an action only one matters, namely the difference it makes to the promotion of some favoured global value. Saying that for consequentialists only consequences matter in the moral appraisal of an action is merely a confused and misleading way of saying that for them only the global point of view matters. All other dimensions of an action, including many which we would ordinarily include among its consequences, are morally irrelevant.

Because consequentialist frameworks attempt to resolve all moral issues from the same ultimate standpoint, they are unlikely to encounter problems of coherence. Instead, the characteristic criticism of such frameworks is that they are narrow or single-minded, blind to the variety and complexity of the moral life. Although this criticism can take many different shapes, the most common objection is to the submergence of the standpoint of separate individuals in the single, dominant, transcendent standpoint furnished by a basic goal.[11] This objection might appear somewhat less damaging if it could be shown that a consequentialist framework is capable of accommodating the separateness of individuals, albeit on a derivative rather than a basic level. We can take one step toward showing this if we can find a place within such a framework for genuine moral rights. We now have on hand all of the ingredients necessary for this task. A moral right, we have said, is a

[10] Finnis 1980, 111 ff.; 1983, 86 ff.
[11] In Rawls's well-known formulation: 'Utilitarianism does not take seriously the distinction between persons' (Rawls 1971, 27).

conventional right for which there is a moral justification. A substantive theory of rights must add to this conceptual analysis an account of how conventional rights are to be justified. Consequentialism can supply such an account: a conventional right is (strongly) justified just in case the policy of recognizing it in the appropriate rule system will better promote some favoured consequentialist goal than will any alternative social policy. Different consequentialist theories will, of course, insert different substantive goals into this schema. But the general form of the schema is common to all of them. A basic goal thus appears to be capable of providing the moral point of view from which conventional rights may be justified, thus of imposing an external control on the proliferation of moral rights.

The solution to our problem of finding existence conditions for rights is doubtless not as straightforward as this simple story makes it appear. In the next section we will deal with some of the more formidable theoretical obstacles which it must surmount. Meanwhile, however, it should already be apparent why consequentialists are able to evade the dilemma encountered by contractarians. That dilemma resulted from the fact that deriving conventional rights from a merely hypothetical agreement with no prior moral controls does not seem to provide them with a moral justification, while imposing any such controls appears to be inconsistent with the contractarian's subjectivist methodology. There is clearly no analogue to this dilemma for consequentialists. Because their theory of value consists of a set of objective, agent-neutral goods, their justificatory framework embodies from the beginning the impartiality which is characteristic of the moral point of view. If a reasonable case can be made for a particular set of goods, and for a particular way of combining them into a global goal, then we will have no further difficulty in understanding why showing that some conventional rights promote this goal should count as a moral justification of them. Of course, the whole enterprise of consructing a consequentialist theory of rights rests on being able to take these first steps. Consequentialists thus owe us both an objective theory of value and its companion theory of rationality. But the present issue is not whether they can deliver these resources, thus not whether they can succeed in their own terms. The problem with contractarian theories was not that they could not succeed in their own terms but that even if they did so the results seemed devoid of moral significance. Although the consequentialist enterprise is scarcely free of hazards, this is not one of them.

6.2 GOAL-BASED CONSTRAINTS

There are many possible consequentialist goals, thus many possible varieties of consequentialism. Since our question is whether it is possible to accommodate rights within a consequentialist framework, it is important for us to abstract from the issues which individuate these varieties. At the same time, however, it is rather fatiguing always to keep the full range of alternatives in play. The business of this section will therefore be greatly expedited if you agree to accept the following offer. You select your favourite goal. It can begin with any menu of ultimate goods you wish, whether of individuals, collectivities, or both, as long as their value is both objective and agent-neutral. And it can combine these separate goods into a single global value in any way you wish, whether aggregatively, distributively, or some blend of the two. Thus your favourite global value might be the general welfare, or the satisfaction of everyone's basic needs, or equality of resources, or peace on earth, or the stability and integrity of the ecosystem, or whatever. The only assumption I will make, purely for expository convenience, is that your goal consists of maximizing this global value. Assuming the necessary kinds and degrees of commensurability, your goal will therefore determine a moral ordering of any set of alternatives, the best alternative being the one which best promotes your preferred global value.

In the arguments to follow, I want you to keep in mind that we are talking about the very best form of consequentialism which you have been able to devise. The question then is whether that form is capable of providing a foundation for moral rights. It might seem odd that I am still treating this question as an open one. After all, have we not already satisfied ourselves that a consequentialist goal provides a method of moral justification? What then could stand in the way of combining your favourite goal with our analysis of a moral right to yield a substantive theory of rights? But in fact something may stand in the way: while a consequentialist goal may be quite capable of justifying some things it may also, by its very nature, be incapable of justifying rights.

One could easily be led to this conclusion by the functional differences between rights and goals. Since goals are fresher in our minds at this stage, let us start with them. A goal contains a global value, which is the resultant of some combinatory operation on some set of local values. Because the maximization of this global value is morally decisive in a consequentialist framework, local losses may be tolerated for the sake of net overall gains. The justificatory procedure appropriate to the consequentialist's transcendent point of view is thus that of a cost/benefit analysis, in which losses are balanced against gains in order to secure the most favourable overall outcome. It is therefore of the essence of a goal that each of the local goods which comprise it is itself dispensable or replaceable; whether it is to be promoted or protected will be determined by what is best on the whole.

By contrast, the function of rights is to constrain the pursuit of such goals. This function can be initially characterized in the following rough way: you take a right seriously only if, at least sometimes, you refuse to violate it even when doing so would better serve your favourite goal. Rights thus confer upon their holders some measure of security against the demands of the impersonal point of view, by ensuring that they will not be routinely sacrificed for the sake of a more favourable overall outcome. Our conception of rights as protected choices reveals the ways in which they perform this function. Every right determines some domain over which its holder is assigned discretionary control. In the case of a liberty-right the domain is given by the bilateral liberty which is the core of the right, whereas in the case of a claim-right it is given by the peripheral powers and liberties which enable the right-holder to manipulate the core claim. In either case the right-holder's autonomy over that domain is secured by imposing on others an appropriate set of duties and disabilities, thereby limiting what it is permissible or possible for them to do by way of compromising that autonomy. Every right, therefore, may be viewed either from the standpoint of its subjects, whose autonomy it enhances, or from the standpoint of its objects, whose autonomy it confines. A right confers on its subjects what has been called an agent-centred prerogative: a moral licence to embark on some course of action even if doing so will not be for the best, as determined from the impersonal point of view.[12] Contrariwise, it imposes on its objects

---

[12] Accounts of the nature and function of agent-centred prerogatives and

an agent-centred restriction: a moral prohibition against embarking on some course of action even if doing so would be for the best, again as determined from the impersonal point of view.

Rights and goals thus appear to have incompatible normative functions. A goal is by definition agent-neutral, assigning to all agents the common aim of pursuing it as best they can. A right is by definition agent-relative, assigning exemptions from pursuit of the goal to some agents and constraints on pursuit of the goal to others. But then it is obvious why the project of deriving rights from goals is in jeopardy. If a right is to be grounded in a goal then the goal must justify constraints on its own pursuit. But surely if we once adopt a goal then we are committed to doing on every occasion whatever will best achieve it, in which case we are committed to ignoring or overriding any such constraints. A basic goal thus appears to be incapable of generating any genuine restrictions on its own pursuit. But then a goal-based moral framework is incapable of justifying any genuine rights.

The problem of reconciling commitment to a basic goal with acceptance of constraints on its pursuit is surely the greatest obstacle in the way of constructing a consequentialist theory of rights, and the principal reason that such a theory has been widely regarded as unworkable. The problem is sufficiently serious that if no solution can be found to it then the whole enterprise must simply be abandoned. Despair, however, would be premature at this point, since it remains to be seen whether we can assemble such a solution out of the resources of a consequentialist framework. The problem appears to be insurmountable because consequentialists appear to be committed to a very simple linear relationship between their moral theory and their moral practice. Imagine, once again, that you have adopted your favourite goal as the sole basis of your moral framework. Then your ultimate aim will be to do on each occasion whatever will best promote that goal. Now suppose we ask how you are to succeed in achieving this aim. The answer seems obvious: among the various alternatives available on each occasion you should try to identify the one which will best promote your basic goal. Thus on each occasion you should conduct a cost/benefit comparison of the available alternatives and you should always choose the course of action which promises the most favourable

restrictions may be found in Scheffler 1982; Slote 1985, chs. 1 and 2; and Nagel 1986, ch. 9.

balance of benefits over costs, and thus the best net outcome. If this is your moral decision-making procedure then clearly you will never have any reason to conform to a genuine constraint on the pursuit of your basic goal. There are, after all, only two possibilities: either the action required by the constraint is the best one available to you or it is not. If it is the best available then the constraint is merely apparent, while if it is not the best available then conformity to the constraint will be counterproductive. In either case the constraint appears to be irrelevant to your moral deliberations.

Now it seems to me that this is a very attractive picture of consequentialist decision-making, and that its attractiveness has always been the main source of the enormous appeal of act-consequentialism. The picture is attractive in part precisely because it seems so truistic: once we have adopted a basic goal, how else could our moral practice possibly proceed? In order to make the picture seem a little less inevitable we need to examine the basic level of a consequentialist moral structure a little more closely. We are assuming that this level consists of a single principle which tells you that the best thing to do is always whatever will best promote your favoured goal. But what is the function of this principle? What does it do? What is it for? On the realist versions of consequentialism with which we are working the answer seems obvious: together with the necessary empirical information, it is supposed to yield a moral truth, or disclose a moral fact, about your situation. And it is supposed to do this by linking a certain causal property (maximizing the promotion of your favourite goal) with a certain moral property (being best), so that the course of action which possesses the first property also possesses the second. Now the members of a feasible set are both mutually exclusive and jointly exhaustive; thus exactly one of them will actually occur while all of the others will remain merely unrealized possibilities. There is an obvious respect, therefore, in which the property of maximizing the promotion of your favoured goal is counterfactual: the maximizing alternative is the one which, if chosen, will better promote your goal than would any of the competing alternatives. Your basic principle then states that the alternative available to you which really is the best is the one which really has this complex counterfactual property.

There is an obvious sense, therefore, in which the perspective adopted by your basic principle is not that of you embedded in a deliberative context but that of an infallible observer. That is, the principle specifies what it would really be best for you to choose, given the way the world actually is, not what you might reasonably choose, given the information available to you. It thus yields the right answer to the question of what you should do, whether or not you have any means of discovering this answer. In ordinary choice contexts it is a commonplace that the option which appears best at the time of decision may turn out not to be best. We can give this slippage a slightly paradoxical flavour by saying that one and the same action may be both the right one to have chosen and the wrong one to have done, but the two assessments are consistent by virtue of being made from different evidential standpoints. A consequentialist framework begins with a principle which operates from the ideal evidential standpoint—the perspective from which all relevant facts about the world are visible. This is why it tells you what really is the case, not what might reasonably appear to you to be the case given your imperfect evidential position.

The basic principles of most moral frameworks, perhaps of all, take this omniscient, after-the-fact, as-things-turn-out point of view. Because different frameworks operate with different basic moral categories, we need some characterization of this perspective which abstracts from all such categories. Let us say that the function of a framework's basic principles is to determine when actions (as well as other aspects of agency) are justified, thus that these principles constitute a theory of justification. Then the principle that an action is best just in case it best promotes your favoured goal is (the sole ingredient of) a theory of justification. This theory, when equipped with the information it demands, gives the right answer to all moral questions. What it does not do, however, is tell you how to discover that answer. While it specifies the target you are to hit, it gives you no strategy for hitting it. If a moral theory is to provide a guide to action, its theory of justification must therefore be supplemented by what we may call a theory of decision-making. This additional ingredient will then give you a procedure or practical policy to follow when confronted by moral problems. Whatever shape this procedure might take, unlike a theory of justification it will need to take into account the various contingencies which are in play when you must deal with such problems: your abilities, your conscientiousness, the pressures of your situation, the extent and reliability of the information available to you, and so on. Thus whereas a theory of moral justification takes the perspective of an omniscient observer, a theory of moral decision-making takes the perspective of a real-life moral agent.

A complete moral framework, and thus a complete consequentialist framework, must include both a theory of justification and a theory of decision-making. We know what your theory of justification is, but what is your theory of decision-making? We are now in a position to see that the simple picture sketched earlier represents one possible answer to this question. After all, the most obvious strategy for finding the maximizing alternative on each occasion of choice is simply to seek the maximizing alternative on each occasion of choice. On this view of the matter your theory of moral decision-making merely replicates your theory of moral justification. Because on this strategy you try to satisfy your basic principle by simply trying to satisfy your basic principle, let us call it the straightforward or direct strategy.[13]

This picture of a consequentialist decision strategy has some obvious merits. For one thing, since your basic principle does double duty as your practical policy it is agreeably economical of theoretical resources. For another, since on this account all of your theory's practical implications are deducible from its basic principle plus empirical truths about the world, there is no possibility of genuine inconsistency between theory and practice. However, if we are now in a position to see that this is one possible theory of decision-making for a consequentialist, we are also in a position to see that it is not the only one. Consider the following question: how are you to select among competing decision-making strategies? Your only basis for evaluating them appears to be your theory's basic principle. That principle tells you that the alternative you should choose on each occasion is the maximizing alternative. It therefore furnishes the standard for assessing competing strategies: one strategy is superior to another just in case it will enable you to identify the maximizing alternative more reliably on occasions of

---

[13] There have been several treatments in recent years of the distinction between theories of justification and theories of decision-making, and of the companion distinction between direct and indirect decision-making strategies: Williams 1973, sect. 6; Hare 1981, chs. 2 and 3; Scheffler 1982, 42 ff.; Parfit 1984, ch. 1. Historical precedent for the adoption of an indirect strategy may be found in Mill 1969, 224–6, and in Sidgwick 1962, 413, 423 ff. The contractarian analogue of an indirect consequentialist strategy is defended in Gauthier 1986, ch. 6.

choice. Or in other words, the best strategy is the most successful, the standard of success being given by your theory of justification. Now one possibility is that you will succeed most often at choosing the maximizing alternative if you simply aim on each occasion at choosing the maximizing alternative. This is the hypothesis which, if true, would support the direct or straightforward strategy. But it might not be true. In any case, it is an empirical hypothesis. Thus the fact that the straightforward strategy is straightforward—that is, that it replicates on the practical level your theory's basic principle—is itself no point in its favour. The basis for choosing among competing strategies is not theoretical symmetry or elegance but success rate. And the most successful strategy might not be the most direct strategy.

Any indirect strategy will consist of some policy other than, or in addition to, just aiming at doing your best on each occasion. It will therefore propose some more complex set of deliberative procedures or dispositions designed to enable you to satisfy your basic principle more reliably. Clearly there are a great many possible indirect strategies. What we now know is that we should favour the direct strategy over all of its indirect competitors just in case its success rate promises to be higher. We have accomplished something just by showing that it is not obvious on the face of it that the direct strategy is the one to choose. But we need to go further. Since we want to know whether a consequentialist goal can justify genuine constraints on its own pursuit, we need to determine whether the best decision-making strategy is likely to include such constraints. Could such a strategy ever either permit or require you not to choose what appears on the best information available to you to be the maximizing alternative? If so, then such a strategy would include respect for rights both as agent-centred prerogatives and as agent-centred restrictions, and a consequentialist moral structure could be capable of building into its rules of practice the very constraints which are alien to its basic principle. The programme of constructing a consequentialist theory of rights thus depends on making a case for the appropriate sort of indirect or constrained strategy.[14]

Since the relative merits of the two theories of decision-making

---

[14] For recent discussions favourable to such a case see Brandt 1984, Gibbard 1984, Gray 1984, and Griffin 1984. For a defence of the contrary view see Frey 1984.

rest on very complex empirical considerations, the best we will be able to manage is a presumptive case in favour of a constrained strategy. I will proceed by dealing separately with two different contexts in which a constrained strategy might seem attractive to a consequentialist. The first context is that in which a society collectively constrains its own pursuit of its own favoured goal, while the second is that in which an individual agent does so. If a case can be made for such self-imposed constraints in both contexts, then we may conclude that both societies and individuals can have consequentialist reasons for recognizing and respecting conventional rights. First, then, the social context. A society imposes constraints on itself if it incorporates into its rule system provisions which sometimes render the straightforward pursuit of its favoured goal either impossible or impermissible. The only such rules which need interest us are those which confer rights. The status of these rules may vary from a constitutional charter of rights through common-law principles to ordinary rules of the system; what they all have in common is that conformity to them on the part of public officials will impede their direct pursuit of the socially favoured goal by the means which they deem most efficient. The officials are therefore constrained not to adopt some policies or courses of action even though they believe quite reasonably that these would best promote that goal. In trying to locate a goal-based rationale for such self-imposed constraints we could in principle focus on any form of institution or association capable both of collectively pursuing some goal and of recognizing some set of rights. However, if we choose to deal with the more complex forms of institution, such as the state and its network of legal rights, then the empirical factors bearing on the choice of a strategy will rapidly become unmanageable in the space available to us. Instead, therefore, I will tell a more modest story.

Like most universities, my university sponsors a good deal of biomedical and social-scientific research which utilizes human subjects. Again like most universities, my university requires all such research to be submitted to ethical review. As a member of our review committee I regularly assess experimental protocols by means of guidelines which impose two different requirements: (1) that the experiment promise to yield a satisfactory overall ratio of benefits to costs; and (2) that it provide adequate protection for its subjects. A protocol is accepted only if it satisfies both requirements.

In distinguishing these two requirements the practice of my committee is congruent with that of similar committees in other institutions.[15] The first requirement requires a cost/benefit balancing in which the main category of costs consists of the harms to which experimental subjects will be exposed while the main category of benefits consists of the pay-offs, either for the subjects themselves or for society at large, yielded by the results of the experiment. At this stage of deliberation a protocol must promise an acceptable ratio of benefits to costs. Our guidelines define an acceptable ratio in the following way: 'the foreseeable overall benefit of the proposed research to science, scholarship, understanding, etc., and to the subjects, must significantly outweigh the foreseeable risks subjects may be invited to take'.[16] A protocol whose cost/benefit ratio is deemed to be unacceptable will be given no further consideration.

Once it has cleared this first hurdle, a protocol must then satisfy a number of further conditions designed to protect the well-being and autonomy of research subjects. Some of these conditions govern such matters as confidentiality, remuneration, and the protection of special categories of subject, but the most prominent and most stringent of them is the requirement that subjects be adequately informed of the nature and purpose of the experiment, including its anticipated risks and benefits, that they should consent to participation freely (without coercion or duress), and that they should be able to terminate their participation whenever they wish (without a penalty of any kind). Our guidelines treat this requirement of informed consent as a way of safeguarding the dignity or autonomy of the subjects: 'The primary reason for requiring consent is the ethical principle that all persons must be allowed to make decisions and to exercise choice on matters which affect them.'[17] It safeguards the subjects' autonomy by conferring on them the liberty-right to participate or not as they choose. Considered as a whole, therefore, my committee's guidelines operationalize the view that, however favourable an experiment's cost/benefit ratio might be, it is unacceptable if it unjustifiably violates the rights of its subjects.

The requirement of a satisfactory cost/benefit ratio formulates a goal, while the requirement of informed consent constrains the pursuit of that goal. In drawing up its guidelines, therefore, my

---

[15] See National Commission 1978, Part C; Working Group 1978, ch. 4.
[16] Dickens 1979, 21.   [17] Ibid. 24.

committee has elected to limit its freedom to seek on each occasion the most favourable overall balance of benefits over costs. There is in this nothing surprising; as I have said, similar committees elsewhere work with the same constrained policy. But now let us suppose, contrary to fact, that my committee's basic goal is to maximize aggregate welfare on every occasion. The direct or straightforward strategy for achieving this goal would be to drop the second requirement entirely and simply assess each protocol solely on the basis of its expected cost/benefit ratio. From the point of view of this basic goal, therefore, our actual practice can only be an indirect or constrained strategy. What reason might we have for thinking that this constrained strategy will be more successful than the straightforward strategy?

Before attempting to answer this question we should satisfy ourselves that our policy does indeed embody genuine constraints. After all, even adherents of a direct strategy will usually have good reasons for requiring the informed consent of experimental subjects, reasons which stem from the costs which are likely to result from bypassing such consent. These costs are typically of three sorts: (1) the suffering, anxiety, distress, etc. which subjects will undergo during this particular experiment and which they would have declined had they been fully briefed in advance; (2) the erosion of the subjects' sense of self-worth which results from being treated as mere tools in the hands of the researcher; and (3) the weakening of subject protection on future occasions. These considerations suffice to show that even if my committee employed a direct strategy we would still attach considerable importance to securing the informed consent of subjects. None the less, this strategy could not result in assigning the requirement the role which it actually plays in our deliberations. On the direct strategy once all the costs of violating informed consent have been included in the overall cost/benefit equation the moral weight of this consideration has been entirely exhausted. This then leaves open the possibility that these costs might be outweighed by the experimental benefits, so that a protocol which violates informed consent might for all that return a favourable balance of benefits over costs. In our practice, however, the requirement of informed consent plays a different role. In attempting to reckon an experiment's cost/benefit ratio we do, of course, try to include all of its negative effects both for its own subjects and for the subjects of future experiments. In this our practice is identical to the direct strategy. But we then depart from this strategy in two important respects. At the stage of a cost/benefit analysis we may reject a protocol despite the fact that its expected benefits exceed its expected costs. For our standards a marginally favourable balance of benefits over costs is insufficient: the benefits must 'significantly outweigh' the costs. Then, even if a protocol passes this test we may reject it because it violates informed consent. In this we are treating informed consent not just as another item in the cost/benefit balance sheet but as a consideration with an independent weight of its own. Thus we will sometimes reject a protocol because it compromises the integrity of its subjects even though, on the best evidence available to us, we think that it will return an acceptable ratio of benefits to costs. Of this practice the direct strategy can make no sense.

Our indirect strategy is therefore not just a sophisticated or disguised form of act-consequentialism. In any consequentialist moral structure the basic goal determines the theory of justification for actions. Therefore, if act-consequentialism is a distinctive version of consequentialism it must be a theory of decision-making rather than a theory of justification. But then it must consist in the injunction to pursue one's favoured basic goal by means of a direct strategy. Since we treat informed consent as an independent constraint, we do not base our decisions solely on consideration of what will be for the best, as measured by our (supposed) welfare-maximizing goal. But in that case we do not employ a direct decision-making strategy, and so we are not behaving as act-consequentialists. Instead, we are basing our decisions both on the direct pursuit of our basic goal and on respect for the subjects' right of informed consent.

Because our strategy contains these two distinct factors it is not rule-consequentialist either. In any consequentialist moral structure the basic goal also determines the theory of justification for rules. Therefore, if rule-consequentialism is a distinctive version of consequentialism it too must be a theory of decision-making. But then it must consist in something like the injunction to base one's decisions solely on the rule general conformity to which will best promote one's favoured basic goal. Our practice violates this injunction in two different ways. First, we do not employ our procedure because we think that it would be for the best if all review committees conformed to it; instead, we employ it because we think it will be for the best if we employ it. Secondly, to the extent that our procedure embodies a rule—the requirement of informed consent—we do not decide particular cases solely by reference to that rule. Furthermore, we do not treat the rule as indefeasible by direct appeals to our goal. Instead, we allow it to be overridden in two different sorts of case. Where subjects are competent to give or withhold consent we permit a protocol to incorporate some deception, as long as (a) the subjects are fully debriefed afterwards, and (b) they are exposed to no more than a negligible risk. And where subjects are not competent to consent we permit investigators to secure proxy consent, as long as (a) the subjects themselves stand to benefit from the results of the experiment, or (b) they are exposed to no more than a negligible risk.[18] In both sorts of case our practice is again similar to that of other committees.[19] However, in both we are allowing some trade-offs between our two requirements, thus balancing direct pursuit of our goal against conformity to our constraint.

The net effect of our complex decision-making procedure is to confer upon research subjects a defeasible right to informed consent. Because this right operates as a genuine constraint, it raises a threshold against acting solely on the basis of a cost/benefit balancing. But because the right is defeasible, this threshold is not insurmountable. Where its balance of expected benefits over expected costs fails to surmount this threshold we will reject an experiment because it proposes to bypass informed consent. But where this balance does exceed the threshold, and where the costs are negligible, we may accept an experiment even though it proposes to bypass consent. Our strategy is therefore genuinely constrained, but also sensitive to straightforward considerations. The remaining question is: why might such a complex strategy be the best means of achieving our basic goal?

I propose to turn this question around and ask instead what the best case for a direct strategy would be. After all, it is possible that my committee's particular procedure is not the best means we could devise of achieving our own basic goal. Even if this is so the case for goal-based constraints need not be jeopardized, as long as some constrained strategy would be best. The indirect strategist is arguing the relatively weak thesis that some constrained strategy or

[18] The foregoing is a greatly simplified version of Dickens 1979, 28–30, 34–6.
[19] See National Commission 1978, 12–13; Working Group 1978, 23–4, 30–1.

other will be more successful than a straightforward strategy. The direct strategist, by contrast, must establish the much stronger thesis that a straightforward strategy will be more successful than any constrained strategy. What conditions would have to be satisfied by our decision context in order for this thesis to be plausible? With no pretension to completeness, I can think of three such conditions.

1. *Unlimited domain of options.* If there are no predetermined or externally imposed constraints on the set of available alternatives then, in principle at least, new options can always be devised, and old ones further refined, in order to yield better outcomes. All that is fixed for us is our maximizing goal; we are free to pursue any pathway, whatever it may be, which promises to get us to it efficiently. This ability to manipulate and expand the feasible set results in turn from our control over the agenda. Where someone else determines the available alternatives, and the order in which they are to be considered, we may be able to do little more than accept or reject each as it comes along. By contrast, where we are the initiators we are free to present new alternatives at any time and to adopt any procedure we like for deciding among them.

2. *Perfect information-gathering.* Of course, the absolutely best case would be decision-making under certainty about the outcomes of all options. This would require omniscience. But we could get by with something less, namely decision-making under risk with reliable objective probabilities. In either case, the process of acquiring information must itself be costless.

3. *Perfect information-processing.* Assume that the goal of maximizing aggregate welfare is itself coherent and determinate. Then once we have reliable information about the individual gains and losses which will result from each alternative, we still need to generate overall cost/benefit ratios for each. This will require both freedom from bias and formidable computational skills. Furthermore, information processing, like information gathering, must itself be costless.

If we combine these three conditions we get a profile of the ideal agent for a direct strategy: someone who is extremely powerful, highly knowledgeable, exceptionally bright, and rigorously impartial.

Does this remind you of anyone you know? If I were such a creature then I would be pretty confident that I could identify the maximizing option in most decision contexts. I would therefore regard a direct strategy as my best bet. (I leave aside here complications for the direct strategy which may be caused by co-ordination problems.) Since a direct strategy involves just directly trying to satisfy one's basic moral principle, this result should not be surprising. For we have already observed that this principle, by virtue of determining the right answer regardless of whether we have access to it, takes the perspective of an omniscient observer. If such an observer is also omnipotent, hyperrational, and impersonally benevolent then he/she is ideally equipped to hit the target of maximizing welfare by aiming directly at it.

It would be tedious to dwell now on the many respects in which our world falls short of this decision-theoretic utopia. I will therefore confine myself to contrasting with it the decision context in which my review committee operates. First, we have only a limited degree of agenda control. We are not ourselves the initiators of the proposals which come before us. Thus our options are pretty well limited to accepting or rejecting a proposal. We can, of course, impose (or negotiate with the investigator) conditions which must be satisfied if a protocol is to be accepted. But beyond this we have little power to amend or redesign a protocol. Furthermore, we are confined to considering those proposals that happen to come along. The process by which research projects are generated in the first place is sensitive to a large number of contingencies: the career pressures to which researchers are exposed, the dominant ideology within a scientific domain, the structure and funding of research institutions, the vagaries of public demand and political agendas, the priorities of granting agencies, and so on. We are free only to decide whether the proposals which are regarded as scientifically respectable, by the prevailing standards of the discipline, are also ethically acceptable. We have no power to suggest that an entire line of inquiry has become bankrupt, nor that research priorities should be fundamentally rethought, nor that the funds which are being poured into research could be better utilized in some other fashion.

Indeed, one feature of our review committee, which it shares with most committees elsewhere, serves to guarantee that these deeper issues will seldom be examined. The committee responsible for assessing a particular protocol will normally contain a majority of members who are themselves researchers in the domain in question, though not connected in any direct way with the proposal under consideration. The rationale behind this arrangement is that only experts in the field can determine the scientific merits of a protocol. While this is doubtless true, the upshot is that the process of scientific review basically consists in determining whether the protocol satisfies the standards of experimental design which are currently accepted in the domain in question, and which therefore are likely to be common ground between the initiator of the proposal and its scientific assessors. By the very nature of the case, challenges to the viability of an entire research domain are unlikely to be raised, or to be taken seriously if they are raised.

In the second place, our access to information is as limited as the scope of our powers. If we were to conduct a full cost/benefit analysis we would need reliable information concerning both the benefits promised by the research and the risks to which subjects will be exposed. For both sorts of information we are largely dependent on the initiator of the proposal, who of course is a committed advocate rather than an impartial consultant. The danger here is not generally deliberate distortion but rather selective perception. Since investigators should be expected to believe in the validity of their own proposals, it would not be surprising if they tended to overstate the importance of the expected results and understate the risks to the subjects. On the benefit side, a protocol will of course identify the broad area of inquiry and describe the contribution to it intended by this particular experiment. These are claims which the scientific members of our committee should be in a position to assess, but for the reasons given above their very expertise in the area is likely to mean that they share the investigator's belief in the importance of the expected outcome. In practice, therefore, their role is largely confined to checking the validity of the investigator's planned procedure. Furthermore, once our committee has passed a protocol we have little or no further involvement in the experiment. While investigators are mandated to report any significant departures from the approved protocol, lay members of the committee have no real opportunity to determine whether the benefits so confidently predicted beforehand are ever realized after the fact. Our acceptance of these predictions, typically advanced by investigator and scientific assessors alike, is largely an act of faith. Meanwhile on the cost side, we are dependent on risk assessments which are furnished by the party who will be imposing them rather than the parties who will be exposed to them. Since the investigator is the initiator of the proposal, there is an institutional channel through which that side of the case can be made. By contrast, the subjects, who will only be recruited once the protocol has been approved, are in the nature of the case unorganized and unrepresented. In principle this structural bias in favour of the investigator could be mitigated somewhat by appointing a counterbalancing advocate of the subjects's interests. But our committee has no such advocate.

Finally, those of us who have the task of processing all of this imperfect information bring to it all of our own idiosyncrasies and prior commitments. When we attempt to compare a protocol's expected benefits and costs we make no pretence of using a precise metric. Instead, we work with highly impressionistic labels, such as 'significant' and 'negligible', behind which it is easy to conceal the influence of one's preconceptions. Furthermore, the deadlines under which we operate and the outside pressures imposed by our other duties conspire to limit the duration of our deliberations, and thus the extent both of our information-gathering and our information-processing. In the end the cost/benefit ratio which we project for a protocol is much more likely to be a matter of intuitive guesswork than the product of a rigorous quantitative analysis.

For all of these reasons my committee is a far cry from the ideal utilitarian administrator with unlimited opportunities to channel social energies down optimal paths. Were we such an administrator then we would indeed base our decisions solely on the outcome of a cost/benefit balancing. The practical limitations of our decision context, however, induce us to depart from this direct strategy in two distinct ways. The first is a retreat from maximizing to something much more like satisficing. Since in evaluating a protocol my committee's options are pretty well confined to acceptance or rejection, our concern is not with whether it is the best possible research initiative, or the best possible use of the proposed funding, but only with whether it passes some absolute standard. Furthermore, the defects in our cost/benefit information, plus the fact that risks to subjects are generally more predictable than experimental benefits, lead us to set this standard fairly high. Thus my committee's first test, namely that a protocol's expected benefits must 'significantly outweigh' its expected costs.

The second departure is toward what has been called pre-commitment.[20] Pre-commitment is an indirect strategy for coping with weakness of will or limitation of resolve. It is indicated when you have antecedently chosen to pursue some goal but have reason to fear that, at the moment of decision, you will blunder down the path which frustrates that pursuit. Roughly speaking, to pre-commit is to increase the likelihood of choosing the antecedently preferred option when the particular occasions for choice arise, by manipulating one's environment so as to reduce either the feasibility or the desirability of the competing options. Although it can take a number of forms, two of the most common involve making the seductive option either physically impossible or much less desirable. Classic cases include the strategies of weak-willed smokers who try to boost their chances of quitting either by placing themselves in situations in which tobacco will be unavailable or by licensing their friends to ridicule them in the event of their backsliding.

The many defects of the decision context faced by my review committee render pre-commitment an attractive option. Being aware of these defects, we know that if we attempt a full cost/benefit analysis of each experimental protocol we will very often make mistakes, thereby permitting unnecessary costs to be imposed on research subjects. Furthermore, the probability of these costly mistakes will remain very high even after we have shifted from the comparative/maximizing to the absolute/satisficing version of the cost/benefit test. However, being also aware of our commitment to the goal of maximizing welfare, we have reason to fear that the temptation to make the attempt might be irresistible on each particular occasion. In order to defeat this temptation we will do well to pre-commit ourselves by announcing from the outset a requirement of acceptability for protocols whose function is to constrain acting on the basis of the cost/benefit test. Thus our second test, namely that however favourable a protocol's cost/benefit ration might appear to be, on the best evidence available, it must also satisfy the further requirement of informed consent. By adopting this constraint we oblige ourselves not to treat informed

---

[20] In Elster 1984, ch. 2.

consent as just another entry in the cost/benefit balance sheet, thereby committing ourselves to rejecting some protocols which promise to yield an acceptable cost/benefit ratio. We also effectively decentralize risk assessment, thereby counterbalancing our reliance on the investigator's perception of risk by giving veto power to the prospective subjects.

Our rationale for endorsing this pre-commitment, in advance of considering particular cases, rests largely on our lack of confidence that we will be able to project costs and benefits accurately in those cases. At the stage when we are designing the standards which we will later be committed to applying in particular cases we can choose to treat informed consent either as an item in the cost/benefit balancing or as an independent constraint. Because of the many impediments which will afflict our decision context on those later occasions, our judgement is that if we allow informed consent to be violated whenever doing so seems (on balance) beneficial then we will often allow it to be violated when doing so is in fact (on balance) harmful. If this judgement is correct then we will better achieve our goal in the long run if we treat informed consent as an independent constraint, thus if we do not assess protocols exclusively on the basis of their expected cost/benefit ratios.

Our confidence that we will enjoy a better success rate in the long run with our constrained strategy therefore depends heavily on our estimate of our own fallibility. Whichever strategy we choose, we know that we will make mistakes. For the sake of the argument, suppose that the options open to us on each occasion are simply to accept or reject the protocol before us. Our theory of justification tells us that a protocol should be accepted just in case its resultant cost/benefit balance will actually turn out to be positive. It is therefore possible for us to make two different kinds of mistake: (1) we can accept a protocol whose balance turns out to be negative, and (2) we can reject a protocol whose balance would have turned out to be positive. Call the first kind of mistake a *false positive* and the second a *false negative*.[21] What we need to estimate for each strategy is the likely frequency and seriousness of each kind of mistake. In making such a comparison we will want to focus on those cases in which our choice of a strategy will make a difference

---

[21] I owe the distinction between these two kinds of mistake to Jim Child.

by leading us to divergent decisions. These are the cases in which on the best evidence available to us a protocol's projected cost/benefit balance is positive but securing this positive balance requires violating informed consent. In such cases a straightforward strategy will lead us to accept the protocol while our constrained strategy will lead us to reject it.[22] Thus all of our mistakes on the former strategy will be false positives, while on the latter they will all be false negatives. Given the cost/benefit structure of research on human subjects, the costs of a false positive will fall on the subjects of the experiment while the costs of a false negative will fall on the potential beneficiaries of the experiment. There may be no reason for us to think that either sort of cost is inherently more or less serious when it occurs; thus there may be no reason for us to think that false positives are inherently more or less serious than false negatives. However, the deficiencies of our decision-making structure which I outlined earlier are likely to bias our perception of these costs in the experimenter's favour. Thus we will tend systematically to understate the costs of a false positive to the subjects while overstating the costs of a false negative to the potential beneficiaries. Since this tendency will increase the frequency of false positives if we employ a straightforward strategy, we are likely to make fewer mistakes of this sort if we introduce a counterbalancing bias in favour of the subjects. This counterweight consists partly in raising the standard in our cost/benefit test. But it also involves imposing the further constraint of our informed consent test.

However, imposing such a constraint also threatens to increase the frequency of false negatives. The superiority of a constrained strategy is therefore all the more likely if its constraint is defeasible. An absolute constraint will prevent us from authorizing the violation of informed consent even on those occasions when it promises to yield sizeable benefits with only negligible costs. An overridable constraint, by contrast, holds out the promise of enabling us to realize these gains while none the less avoiding the routine sacrifice of individual integrity for the sake of a projected cost/benefit balance which, while positive, is both marginal and highly speculative. I should emphasize again that I am by no means confident that the particular constrained strategy which my committee has chosen is the best means of serving its (imagined)

---

[22] More accurately: our constrained strategy may lead us to reject it. See the next paragraph.

basic goal. But my aim was not to make a conclusive case in favour of this constrained strategy, but only a presumptive case in favour of some such strategy. And of course my story was only a story. But none of its lessons seem to depend either on the particular goal in question or on features which are peculiar to this particular institution. The case in favour of a constrained strategy should therefore be generalizable. It should, that is, apply *mutatis mutandis* to most societies pursuing most consequentialist goals. More specifically, it should apply to the collective pursuit of your favoured goal, whatever that may be. But in that case societies which commit themselves to pursuing your goal will have good reason for conferring and protecting some set of rights.

Earlier I distinguished two contexts in which the options of a direct and an indirect strategy are both available. Having dealt with the social context, it remains only to consider the case of an individual agent. Here the worry is that while a society collectively might have good consequentialist reasons for recognizing a set of rights, individual agents might have no such reasons for respecting those rights on particular occasions. David Lyons has raised this objection against a utilitarian justification of legal rights:

> . . . when legal rights are regarded as justifiable or morally defensible, they are regarded as having moral force. In other words, the idea that legal rights are morally defensible entails the idea of a moral presumption in favor of respecting them, even though it may not be useful to exercise them or may be useful to interfere with them in particular cases. The problem for utilitarianism, then, is whether it can somehow accommodate the moral force of justified legal rights. I argue that it cannot do so satisfactorily. Although there are often utilitarian reasons for respecting justified legal rights, these reasons are not equivalent to the moral force of such rights, because they do not exclude direct utilitarian arguments against exercising such rights or for interfering with them. Specifically, utilitarian arguments for institutional design (the arguments that utilitarians might use in favor of establishing or maintaining certain legal rights) do not logically or morally exclude direct utilitarian arguments concerning the exercise of, or interference with, such rights. As a consequence, evaluation of conduct from a utilitarian standpoint is dominated by direct utilitarian arguments and therefore ignores the moral force of justified legal rights. The utilitarian is committed to ignoring the moral force of those very rights that he is committed to regarding as having moral force by virtue of the fact that he regards them as morally justifiable.[23]

[23] Lyons 1982, 109–10.

As Lyons himself recognizes, this objection would apply with equal force to any consequentialist moral framework. As he doubtless would agree, it would also apply, perhaps with equal force, to any conventional rights. Thus the general question is whether consequentialist agents would have reason to respect those conventional rights which consequentialist societies would have reason to establish. Lyons's negative answer to this question depends largely on his claim that the moral decision-making of such agents on particular occasions will be 'dominated by direct [consequentialist] arguments'. It therefore depends on the claim that the decision-making strategy dictated for such agents by their basic goal will be direct rather than indirect. But is this so?

Having given this issue extensive consideration in the social context, we can afford to be brief here. Imagine, once again, that your basic commitment is to the pursuit of your favourite goal. Exercise of your own conventional rights will permit you some measure of disregard for that goal, but respect for the rights of others will require it of you. It is therefore the conventional rights held by others which are the real impediments to your straightforward maximizing activity. Now consider the subset of such rights which you regard as morally justified because their recognition in the appropriate rule system is the best policy for promoting your favourite goal. If they are so recognized then this fact will be likely to alter the consequences of your available courses of action by virtue of rendering your invasion of the protected domains of others either less feasible or less desirable. Now suppose that you choose to respect these rights when and only when doing so, all things considered, appears to be the best means of promoting your goal. You are then electing a direct strategy. But, as Lyons points out, you are also denying moral force to these rights by denying them any weight of their own against your pursuit of your goal. On the other hand, if you accord them some such weight then you are electing a constrained strategy. What could lead you to prefer such a strategy?

The case in favour of a constrained strategy for individuals parallels in every respect the case in favour of such a strategy for collectivities. Once again it is essentially an argument from fallibility. If you simply set out on each occasion to do the best you can then the limited information available to you, your imperfect ability to process this information, your natural bias in favour of yourself and those closely connected with you, the various pressures to which you are subject in the heat of deliberation—all of these factors and more will lead you to make costly errors. You are therefore likely to do better if you pre-commit yourself to observe some relatively simple rules even when doing so seems, on the best evidence available to you, to disserve your basic goal. When looking for such rules one obvious candidate will be the requirement that you respect those rights whose conventional recognition is the best social policy for promoting that very goal. Thus you are likely to do better if you adopt a constrained strategy, one of whose constraints is an inhibition against violating the morally justified rights of others. This inhibition need not, and doubtless should not, be indefeasible. Thus your decision-making policy need not, and doubtless should not, entirely exclude direct appeals to your basic goal. But your internalized constraint will raise a threshold against such appeals—the 'moral presumption' of which Lyons speaks—so that you will not violate rights to realize merely marginal and speculative gains. Once again the most successful policy promises to be a mixed strategy which balances direct appeal to your favoured goal against conformity to your self-imposed constraint.

Since a constrained strategy accords independent deliberative weight to rights, it seems to acknowledge their moral force. Recall that we have agreed that taking rights seriously requires allowing them to impede pursuit of one's favoured goal. Since an indirect strategy assigns them this function, and since at least a presumptive case can be made in favour of adopting such a strategy, Lyons's worry seems groundless. However, it may have a source which we have not yet considered. A right has moral force, we have said, only when its existence provides a moral reason for action. Where the right is one's own the reason takes the form of a prerogative, where it is someone else's it takes the form of a restriction. But it might be urged that for an agent committed to a consequentialist framework reasons must always flow from the goal which is basic to that framework, thus that it would be irrational for such an agent ever to choose a course of action which appears, on the best evidence available, to disserve that goal. Since one cannot have reason to do what is irrational, so the argument continues, a consequentialist cannot have an independent reason for respecting rights. But in that case a consequentialist cannot acknowledge the moral force of rights.

This argument turns on the claim that once you are committed to a basic goal it is irrational ever to act on a strategy which constrains pursuit of that goal. The question of whether conformity to a constrained policy could be rational for someone who is fundamentally a straightforward maximizer has been much debated.[24] But in fact we need not answer it. If such conformity can itself be rational then, of course, the foregoing argument fails. But even if it is irrational, it is a case of what Derek Parfit has called rational irrationality: conduct which, though on this particular occasion it frustrates your basic goal, is none the less part of the best long-range strategy available for achieving that goal.[25] This slippage between the directly best option and the option required by the best strategy can occur only for agents with our imperfections. But given that we have those imperfections, we know that we will do worse in the long run if on each occasion we simply aim to do our best. In that case attempting always to be rational will itself be irrational, and if you choose to act in a way which disserves your basic goal you have a good reason for doing so. And that is all that is necessary for you to treat the existence of rights as independent reasons for action.

The case which we have made for rights as goal-based constraints, both on the individual and on the social level, rests heavily on the imperfections of our own nature and of our decision-making environment. It may well be true, therefore, that ideal moral agents in an ideal world would have no reason to acknowledge rights as constraints on the pursuit of their favoured goals. Some readers may be dissatisfied with a moral foundation for rights which applies only to our actual world and to those imaginable worlds which share its imperfections. The idea of a theory of rights tied less closely to these features of the human condition remains very alluring. We can console ourselves, however, by reflecting that the imperfections of our world, while doubtless regrettable, seem equally unalterable. In that case we need not fear that a consequentialist foundation of rights will easily become inapplicable or obsolete. Furthermore, we should recall that the real world is the

[24] See Parfit 1984, ch. 1; McClennen 1985; Gauthier 1986, ch. 6.
[25] Parfit 1984, 13.

one in which the integrity of rights is under attack, both from the zealousness of their supporters and the callousness of their enemies. A theory of rights which is capable of providing them with a foundation in that world is not a thing to be scorned.

7

Aspects of a Theory of Rights

EXISTENCE conditions for moral rights require a substantive theory of rights. A consequentialist moral framework appears to be capable, in principle at least, of supporting such a theory. (Presumably it is equally capable of supporting many of the other agent-relative ingredients of common-sense morality.) In this respect it is superior to both of its leading theoretical rivals. Since the natural rights and contractarian traditions are commonly assumed to be friendly to rights, and since consequentialism is just as commonly thought to be hostile to rights, these results are surprising. We should remind ourselves, however, of the limitations of our methodology. We have confined our attention to the three traditions which dominate contemporary moral/political theory. Since there may well be viable options which we have not considered, nothing in our argument suffices to show that a consequentialist structure is the only one capable of supporting a substantive theory of rights. For all we know, there may be many frameworks with this capacity. In order to refute the nihilist, however, one such framework is enough. And this we now have in hand.

Our victory over the nihilist might, for all that, be a hollow one. A consequentialist theory of rights tells us that a right is genuine just in case the social policy of recognizing it in the appropriate rule system is the best means of promoting some favoured goal. Even if there are no technical impediments in the programme of grounding rights in goals, some formidable practical obstacles remain. Moving from our abstract schema to a functioning theory capable of validating rights would require taking two further steps: selecting a favoured goal and showing that the recognition of some particular set of rights will best promote that goal. Given the daunting nature of both these tasks, a consequentialist structure appears to leave ample room, if not for nihilism, at least for scepticism about rights. Unlike the nihilist, the sceptic does not deny that some moral rights are in fact genuine—only that we can ever know which ones these are. Whereas in order to refute the nihilist we need show only that a consequentialist rights theory is feasible in principle, in order to refute the sceptic we must also demonstrate that it is workable in practice.

No such demonstration will be undertaken here. Indeed, to some extent the sceptic's doubts appear to be well grounded, since within a consequentialist framework there can be no direct, infallible access to the truth about rights. It is the promise of just such access—through moral intuition, or the light of reason, or the authority of some deity—which lies behind the appeal of a natural rights theory. Once we have been shaken out of this dream we may find it discouraging that our rights can be established only at the cost of protracted intellectual labour, including the management of complex empirical data. By comparison with the luxury of consulting our favourite oracle this labour is taxing and unglamorous. We should not, however, be intimidated by its difficulty, nor by the likelihood that we will seldom arrive at final and definitive answers. The actual business of validating rights in a particular sector of our lives usually does not require either that we redesign all of our social institutions from the ground up or that we secure unanimous consent to some utopian ideal. Instead, the practical question is generally how we are to go on from here, thus how we might improve the institutions we already have. In order to make a good case for some reform it will usually be enough if we can identify the principal values at stake in the sector in question and strike some seemingly reasonable balance among them. Clearly this process may lead us to different answers at different times and under different circumstances, but this fact does not impugn the best answer we have been able to develop at a given time under given circumstances. When sceptical doubts are flourished concerning the entire enterprise, it is well to remember both that the truth about rights is unlikely to be simple and that our thinking about them, as about all moral matters, is still in its infancy. We have much yet to learn. If a consequentialist framework does not immediately yield all the right answers, perhaps it is enough if it leads us to ask the right questions.

Providing a more decisive reply to the sceptic would require actually constructing a consequentialist theory of rights. Since different versions of consequentialism are individuated by their basic goals, making a case for some particular version would require showing that one goal is superior to all its rivals. It should be clear by now just how demanding this task would be. We would need to begin by defending the intelligibility of a conception of ultimate value which is both objective and agent-neutral and of a conception of rationality which is able to assess ends as well as means. We would next need to show of some particular inventory of such values that it includes all and only those ultimate agent-neutral ends which it is rational, or reasonable for us to pursue. Then, in order to generate the transcendent point of view which is definitive of a consequentialist theory we would also need to defend both some particular formula for combining these separate goods into a global value and some particular function defined on this value. At each of these junctures the aim would be to show that reason directs us down one path to the exclusion of all others. Even if we could manage all this, however, we would not yet h´ve succeeded in laying the sceptic's doubts to rest. Suppose that your favourite goal can be shown to be more reasonable than any other (or any combination of others). This result would be of merely theoretical interest if we could not establish the causal linkage between your goal and the policy of recognizing some particular set of rights. It is sometimes complained that a consequentialist rights theory, however coherent in principle, would be indeterminate in practice because we could never tell which of a number of alternative social policies would best promote its basic goal. If this is true of your favourite goal then arguments from it will never suffice to ground any moral rights.

The force of this argument from indeterminacy will depend on the basic goal around which a consequentialist moral structure is built. Since it is beyond our means here to make a case for some particular goal it is also beyond our means to provide a decisive reply to the sceptic. Leaving the question in this state, however, would be an unsatisfying denouement to our story. I shall therefore devote the remainder of this chapter to hazarding some guesses about the likely shape of a consequentialist theory of rights. I begin with some assumptions about the likely shape of a consequentialist goal. The raw materials for that goal will be an inventory of ultimate goods. Whatever other goods this inventory might include, it seems reasonable to suppose that in one way or another it will acknowledge the value of those states which are the standard sources or components of individual well-being: life, health, liberty, autonomy, sociality, the development and exercise of powers and abilities, and so on. These goods must then be collated into some global value. Whatever other considerations this global value might include, it seems reasonable to suppose that it will be aggregation-sensitive, thus that it will acknowledge the force of increasing the overall extent to which individuals enjoy these central ingredients of their well-being. Some function must then be selected for this global value. Here it seems reasonable to settle on maximization. By this series of peremptory choices we arrive at a goal which consists at least in part of maximizing the sum of individual welfare. It might well consist of more than this, if it admits other particular goods as well or adopts a combinatory rule which is also sensitive to factors other than their aggregation. My assumption is only that it includes this much, not that it includes no more.

With this (possibly partial) goal in mind, we can begin our quick excursion through a consequentialist theory of rights. It will be convenient to organize the tour around the three dimensions of a right: its scope, content, and strength. The scope of a right has two components: its subjects, or holders, and its objects, or those against whom it is held. Since the latter is the more straightforward, we will begin with it. Who, then, are likely to be the objects of the moral rights validated by our goal? The candidates will clearly be limited by the concept of a right. Our earlier analysis revealed that every right, regardless of its content, imposes some bundle of duties and disabilities on those against whom it is held. The particular bundle imposed by a particular right will be determined by the core and periphery of that right. Since a duty limits what is permissible for an agent, its imposition presupposes that the agent is capable of complying with normative constraints. Since a disability limits what is possible for an agent, its imposition presupposes that the agent is capable of exercising normative powers. Being the object of a moral right therefore requires both the capacity to conform one's conduct to normative constraints and the capacity to alter such constraints. If this pair of capacities is included in the notion of agency then only agents can be the objects of moral rights.

This purely conceptual result is unsurprising. We do not ordinarily think that rights can be held against inanimate objects, or plants, or animals, or infants. The development of agency is a condition of the development of responsibility, including responsibility for violating rights. Thus we may still ask: within the class of agents, who are likely to be selected as the objects of the rights validated by our goal? The only adequate answer is that this will depend on the best policy for each particular right. Many morally justified institutional rights are held against the members of some particular social group, as in the case of rights which students have against their teachers. But if many rights are parochial in this dimension others are likely to be universal. In this latter category we should expect to find those rights to non-interference, and perhaps also to positive aid, which are commonly presumed to hold against everyone in general. In the absence of this universality it is difficult to see how rights could properly safeguard the several ingredients of individual well-being which are included in our basic goal.

The concept of a moral right also imposes limits on those who are capable of holding such rights. Unlike the objects of a right, however, the class of logically admissible subjects will depend on one's favoured model of a right. Since we have settled on the model of rights as protected choices, let us begin with it. On this model every right, regardless of its content, confers upon its holder some bundle of liberties, claims, powers, and immunities, the exact composition of which will once again be determined by the core and periphery of the right. As in the case of an object, being the subject of a right will then presuppose both the capacity to comply with normative constraints and the capacity to manipulate those constraints. But in that case the class of logically admissible subjects will be identical to the class of logically admissible objects.

On any plausible analysis of agency the choice model will deny rights, on logical grounds, to inanimate objects, plants, non-human animals, fetuses, infants, young children, and the severely mentally handicapped.[1] Just as agency is gradually achieved by normal human beings during the course of their maturation, so is the companion capacity to be the holder of moral rights. Lest this result should appear to read everyone except rational adults out of the moral domain, we should remind ourselves that on the choice model rights have the very special function of enhancing and

---

[1] Older children and the more mildly handicapped will clearly possess some degree of agency, thus qualifying (logically) for some rights. There can be no conceptual justification, even on the choice model, for excluding all members of these categories from holding rights.

protecting the autonomous management of one's life. Since they can play this role only for beings who are capable of being self-managers, rights constitute a form of moral protection which is simply out of place for other sorts of creature. To say this is not to deny such creatures other forms of protection. Having moral rights is not the same as having moral standing; to think otherwise is to assume that rights exhaust the moral domain. Restricting rights to agents is therefore compatible with extending moral standing to a much wider class of creatures—perhaps to all those who have interests, or a welfare, which can be protected by the imposition of moral constraints.

Although we may not say that a creature which has interests but lacks agency has rights, we may say that it would be a good thing to safeguard its well-being or that we ought to do so. Indeed, our analysis of rights allows us to say something even stronger. Although we have accepted the choice conception of a right we have also accepted the benefit analysis of a relational duty. It is therefore conceptually possible for us to owe duties to any creature capable of being harmed or benefited, and for any such creature to have claims against us. There is no logical barrier to these duties and claims being moral ones, whenever the policy of imposing the former and conferring the latter will best promote our basic goal. Since the ingredient goods in our goal include the leading sources of well-being, and since our global value is sensitive to aggregate well-being, there would seem to be a *prima facie* case for such a policy whenever our conduct threatens to cause substantial harms which are not counterbalanced by equally substantial benefits. In this way, although a theory of rights which adopts the choice model can make no sense of the rights of animals or fetuses or infants or young children or the severely mentally handicapped, it can accomplish essentially the same objective by making them the beneficiaries of our protective duties.

If you are reluctant to eschew the language of rights in these cases then you will need to switch to the interest model. The prominence of rights rhetoric in contemporary debates about the environment, abortion, and the protection of children and the handicapped shows the extent to which this model has eclipsed its more specialized rival in the public arena. From a logical point of view, extending rights to these additional subjects (on the interest model) comes to exactly the same thing as extending them claims (on the choice model). However, from a practical point of view the language of rights is likely to have considerably greater clout in the moral/political market-place than is the more academic language of claims (and their correlative relational duties). This fact may provide consequentialists with a pragmatic reason for favouring the interest conception, against which the theoretical considerations we offered earlier in favour of the choice conception might seem mere fussiness. The option you prefer here is likely to reflect the importance you attach to the enterprise of defining clear theoretical boundaries, and to reserving rights as special-purpose devices whose primary function is to safeguard one central ingredient of well-being, namely autonomy. Whichever choice you make, it will determine only the way in which you characterize the moral protection to be afforded welfare subjects who do not qualify as agents; it will not weaken your commitment to providing them with that protection.

On either conception of a right a consequentialist framework seems likely to support the existence of some natural rights (in the purely extensional sense). A natural right (in this sense) is a right the criterion for possession of which is a natural property. On any theory of rights we should expect many rights to be tied to some specified institutional role, or social status, or voluntary undertaking, while others will be dependent on the particular circumstances of a society. For both these reasons, therefore, many rights will be parochial rather than universal in their range of subjects. But if rights are to be grounded in a goal whose ingredients include the most prominent components of well-being, then we should expect the rights which are standardly necessary for protecting and promoting these goods to be extended indifferently to all. As we have seen, the limits of this extension will vary with one's model of a right. However, both agency and welfare are natural capacities. Thus on both models some rights are likely to be both natural and universal, in the sense of belonging to all and only those who display the requisite natural capacity.

On the other hand, neither model is likely to support the existence of human rights, if these are rights whose criterion is the natural property of belonging to our species. On the choice model all non-human beings will lack rights if they lack agency, but so will many human beings (fetuses, infants, and the severely mentally handicapped). On the interest model all human beings will have rights if they have interests, but so will many non-human beings (at least some animals). On both models it is quite inconceivable that the extension of any right should coincide exactly with the boundaries of our species. It is thus quite inconceivable that we have any rights simply because we are human. If this is what is implied by the rhetoric of human rights then that rhetoric has been used to serve a discriminatory, because speciesist, programme. However, something weaker than this can be, and often is, meant—namely, that there are some rights which human beings enjoy simply by virtue of being either agents or welfare subjects, therefore not by virtue of any further conventional status. In this more guarded sense some natural rights will also be human rights.

One allegedly human right is the principal point at issue in the contemporary debate over the morality of abortion. Defenders of a conservative view rest their case primarily on the claim that all human beings, and therefore all human fetuses, have a right to life. Although adoption of the model of rights as protected choices makes short work of this claim, it can then be readily translated into an equivalent claim about our duty to protect fetal life. Since the substantive issue is now standardly formulated in terms of rights by both conservatives and their opponents, and since it cannot be resolved by merely conceptual manœuvring, it will be easier for us to address it if we switch temporarily to the interest model. Even on this model the right to life would appear not to belong only to human beings, since life is a central good to many non-human beings as well. But within the abortion debate the important question is whether it belongs to all human beings, including all human fetuses. One's answer to this question will depend in part on one's favoured analysis of interest or welfare. Suppose, as I believe, that welfare is subjective, thus that having a welfare requires being a subject of experiences. It would then seem to follow straightforwardly that a fetus cannot acquire a right to life until it crosses the threshold of sentience, which probably occurs sometime during the second trimester of pregnancy. In this case, since an early (pre-threshold) abortion cannot violate the fetus's right to life it should be regarded as a private matter for the woman who undergoes it.

So I have argued elsewhere, in defence of a moderate view of abortion.[2] This view about how and when fetuses acquire the right

[2] Sumner 1981.

to life seems to have much to recommend it, since it enables us to support the common-sense moral distinction between contraception and early abortion on the one hand and late abortion and infanticide on the other. Furthermore, once we have brought ourselves to adopt the interest model of rights there seems something quite compelling in the idea that a creature can acquire rights only when it has acquired interests which those rights can protect. There is a problem, however, in giving these intuitively attractive results a consequentialist foundation.[3] In the case of natural kinds it is clear why rocks, lacking interests, must also lack rights, while cats, having interests, may also have rights. This is clear in part because rocks, if left to themselves, do not develop into cats. However, a pre-threshold fetus, if left to itself, will develop into a post-threshold fetus. Even before it has actually acquired sentience, and thus a welfare, it has both of these in its future. And for a consequentialist the future must matter. Since a pre-threshold and a post-threshold abortion equally forestall the (further) development of a creature who will (continue to) have interests, how can a consequentialist avoid locating them in the same moral category? But then how can it be, on a consequentialist theory, that fetuses acquire a right to life only when they acquire a welfare?

A short way with these problems is to restrict the ingredients of a consequentialist goal to the interests of those welfare subjects who already exist as such at the time of decision. Unluckily, however, any theory which imposes this sort of restriction on its basic goal can be shown to be unacceptable.[4] A consequentialist goal must treat as equally important the equal goods of present and future beings. But in that case the puzzle about the rights of fetuses remains: why don't the future interests of a pre-threshold fetus count for as much as the present interests of a post-threshold fetus? The solution to this puzzle, if there is one, lies in the resources of an indirect version of consequentialism with its distinction between appeals to a basic goal on the one hand and appeals to rights on the other. If we ask whether a woman's having an abortion is the best option open to her from the impersonal point of view then the answer may be much the same whether the abortion will be

---

[3] See Sumner 1981, ch. 6; Soles 1985; Sumner 1985.
[4] Sumner 1981, sect. 25; Parfit 1984, sects. 134–6.

performed before or after the threshold.[5] In this case whether the fetus has yet become sentient seems to make little or no difference, since it seems to make little or no difference whether the abortion prevents the initial acquisition of sentience or its further development.[6] But in deciding whether to permit women access to abortion we do not want to know whether, or when, having an abortion would be for the best; instead, we want to know whether, or when, having an abortion is within a woman's rights. Since having a right entails having the prerogative not to do what is for the best, this is a different question. Nothing could prevent an abortion from being within a woman's rights except the competing right to life of the fetus. Thus if a fetus has no right to life before the threshold stage and some such right thereafter, a pre-threshold abortion might well be within a woman's rights whereas a post-threshold abortion would not. In a consequentialist framework the distribution of a particular right is a policy question. Whether the threshold makes a morally significant difference in such a framework therefore depends on whether the best policy will be to assign fetuses a right to life when and only when they have become sentient. An argument in favour of such a policy might point to the much greater impact on women of restricting access to early abortions, or to the utility in questions of reproduction of maintaining a distinction between not creating a merely possible person and destroying an actual person, or to the utility in questions of the treatment of animals of maintaining a close connection between sentience and moral protection, or to any number of other global considerations. It is not my present purpose to try to show that the best policy will be to treat the acquisition of sentience as the criterion for having a conventional right to life (though I believe that this can be done). It is enough that it could be the best policy, in which case consequentialists will after all be able to support a moderate view of abortion.[7]

There are a few further issues concerning the scope of rights

[5] Some factors of considerable practical importance, such as the medical risks of abortion for women, do vary roughly with duration of pregnancy. But these factors could not lead us to attribute a different moral status to the fetus at different stages of pregnancy.

[6] It would make a difference, however, if the technique employed would cause a sentient fetus to suffer.

[7] The same policy considerations would also explain why the right to life of a fetus or infant is so strong once it has been acquired.

which we should briefly address. In our analysis of rights we assumed, purely for expository convenience, that the subject and object of every right are distinct individuals. This assumption now needs to be examined. Critics of rights theories sometimes complain that they commit us to an overly atomistic and adversarial picture of social relations. We should therefore ask ourselves whether such a picture will be forced on us if we embrace a substantive theory of rights. Of course we need not, and should not, believe that rights exhaust the universe of moral discourse. Thus even if the language of rights is irretrievably individualist, this tendency could be counterbalanced by other moral categories (such as goals). But is there any reason to assign rights only to individuals? Despite our earlier working assumption, there seems to be no conceptual barrier in the way of assigning moral rights to collectivities. On the choice model, as we have seen, the capacity of agency is a logical pre-condition of having rights. But every social group which qualifies as either an institution or an association must have some procedure for reaching collective decisions and taking collective action. If the individual members of such a group are severally capable of complying with normative constraints and exercising normative powers, there is no apparent justification for denying these capacities to the product of their union. On the choice model, therefore, collectivities will qualify as the subjects of rights as long as they possess the requisite capacity to act on behalf of their members. It is interesting to note that the admission standard under the interest model might be more difficult to satisfy. On that model we would look not for some collective activity which is distinct from the sum of the separate activities of the members but for some collective interest which is distinct from the sum of the separate interests of the members. But whereas the notion of an irreducibly collective agency seems unimpeachable, that of a collective interest which is more than the sum of its parts is problematic. If a collectivity can have a welfare of its own, then the interest model might lead us to attribute rights not only to those groups which are capable of agency but more widely as well. But if not then only the choice model will be able to make sense of collective rights.

However this may be, the concept of a right does not itself lock us into an individualist social/political theory. Once we have cleared away the logical impediment to the idea of collective rights we will be left with the substantive question of which collectivities, if any, actually have such rights. Since in a consequentialist framework this becomes a problem of choosing the most efficient means of promoting a global goal, there is little point here in speculating on the best solution to it. But an exclusively individualist outcome seems on the face of it unlikely, since it would require establishing the strong thesis that our favoured goal will be best promoted by denying rights to all collectivities, regardless of their nature and of our social circumstances. Surely we should instead expect support for a more eclectic social order which acknowledges a mixture of individual and collective rights. There thus seems no impediment, either conceptual or substantive, which could prevent us from appealing to rights in order to ground some form of polity, such as social democracy, which honours both individualist and communitarian values.

There appears to be rather more force in the criticism that a framework of rights, whether individual or collective, encourages us to picture social relations as confrontational. After all, the function of rights is to enable their holders to protect areas of freedom and control against the incursions of others. Since collectivities are capable of colliding both with individuals and with one another, recognizing the existence of collective rights will not alter this function. However, the analytic connection between rights and interpersonal conflict does not justify the allegation that the conceptual framework of rights is appropriate only to the public arena. It is tempting to think that requirements of justice have no place in close personal relations, which are instead the proper domain of an ethic of caring. But to the extent that relations of love and friendship render us more vulnerable to others they also expand and intensify our need for protection against them. Those joined by bonds of intimacy and caring therefore have more, rather than fewer, rights against one another; many forms of conduct which would be supererogatory where strangers are concerned can be demanded of us by our friends.[8] The adversarial presuppositions of a framework of rights will be out of place only in a condition of complete interpersonal harmony. Thus there may well be communities united around both a common goal and a common programme for its pursuit whose members have no need for rights against one another (nor, perhaps, for rules of any kind). For better or worse,

[8] See Badhwar 1985.

however, most intimate relationships do not display this degree of concord.

On the choice model collectivities can be the subjects of rights if they are capable of agency. On any viable model this capacity will qualify them also to be the objects of rights. There is therefore nothing conceptually odd in the notion of rights being held by either individuals or collectivities and also holding against either individuals or collectivities. This leaves us with one final question about the scope of rights: Could the subject and object of a right be identical? Could I have a right against myself? On the interest model there seems nothing incoherent in such a right, since there is nothing incoherent in the idea of imposing a duty on me for my own good. In general, on this model, any rule which imposes a paternalistic duty on me will also grant me a right. (This will of course be a moral right only when the imposition of such a duty can be morally justified.) However, there is something decidedly odd in the idea of my bearing a duty which I have the power to waive or relinquish whenever I please; such a duty seems incapable of serving as a genuine constraint. Thus on the choice model the notion of having a right against myself appears to make no sense. However, if we also accept the benefit analysis of relational duties then we can make sense both of my having claims against myself and of my having duties to myself.

When we turn from the scope of rights to their content there is less that can be said at the abstract level of an overview. The specific rights which will be supported by a consequentialist theory will obviously depend heavily on the basic goal of the theory. Even after this goal has been partially specified so as to include aggregate welfare, the structure of rights derivable from it is far from obvious. The problem is further compounded by the contextual relativity built into the concept of a moral right. As we have already noted, a moral right may be either parochial or universal in scope. Since this is so, we should expect a substantive theory of rights to have two components: a set of universal rights which respond to invariant features of the human condition and a further set of rights which are relativized to the peculiar social, economic, and cultural circumstances of a particular society. In both cases an argument to a particular moral right will need to show that conferring a right with that content on that set of subjects promises to strike the most favourable overall balance among competing interests.

I have no wish to advance any detailed speculations about the likely outcome of this process for any sector of our lives. However, it is easy to imagine how a consequentialist case might be made for many of the standard items in catalogues of human rights. Such a case would point to the importance to individuals of the securities offered both by civil and political rights and (at least where circumstances permit) by social and economic rights as well. If rights in both categories can be supported, with either a universal or a parochial scope, then one broad negative result of a consequentialist rights theory seems inevitable. Recall our earlier distinction between liberty-rights and claim-rights. The difference between them lies in the Hohfeldian position which constitutes their core: a (full) liberty in the first case and a claim in the second. Libertarianism is the view that all natural rights (in the extensional sense) are liberty-rights which impose on others only negative duties of non-interference. There seems to be no prospect of grounding such a view in any plausible consequentialist goal. If it is true, as it surely is, that many liberty-rights can be justified by their tendency to protect the core ingredients of individual well-being, it is also true that many claim-rights—including welfare rights which impose on others positive duties of assistance—can be justified in the same way. It would be quite astounding if the best policy for pursuing our consequentialist goal involved imposing duties which forbid us to interfere with the liberty of others but no duties requiring us to promote their well-being, regardless of how badly off they might be. Although some libertarians have claimed to be consequentialists, most have sensed the futility of this line of argument and have instead sought refuge in a contractarian or natural rights framework. They will presumably be unmoved by the impossibility of giving their structure of rights a consequentialist foundation. But then they will need to supply some alternative.

The final dimension of rights is their strength. Here I will confine myself to a single question: could a consequentialist theory acknowledge any absolute rights? Certainly consequentialist pragmatism will tend to militate against the recognition of any rights utterly indefeasible by direct appeals to a basic goal. However, absolute rights are not an impossible output for a consequentialist methodology. The question we must ask of a putative right is not whether there are any circumstances in which overriding it would be for the best but whether the best policy will be to allow it to be overridden in any circumstances. It may well be true of all rights that their violation will sometimes best promote our basic goal. And it may also be true of most rights that the policy of sometimes permitting their violation will best promote our basic goal. But in other cases we might do better to raise an insurmountable threshold against goal-based considerations even though we think that sometimes these considerations would justify violating the right. This might be so if the exceptional cases are likely to be extremely rare and if allowing ourselves direct appeal to our goal will lead us to violate the right in many non-exceptional cases. Where these conditions are satisfied the additional gains we might realize in the extraordinary cases would be vastly outweighed by the additional losses we would sustain in the ordinary cases, so that we will do better overall to deny ourselves a goal-based override. A plausible instance of this is the claim-right not to be subjected to torture. The case for rendering this right defeasible rests on exceedingly unlikely scenarios in which large numbers of lives could be saved if some vital piece of information were extracted from an unwilling subject. The real practice of torture, however, has nothing to do with these cases. It is quite believable that the price we must pay in order to contain the ordinary practice of torture is to condemn it even in extraordinary cases. If so, then there is a consequentialist case for rendering this right absolute.

This result, like all of the others I have sketched, is of course merely speculative and provisional. More definitive conclusions about the shape of our rights await a full substantive theory. The aim of this book has not been to construct such a theory, but only to clear the ground for its construction. Our point of departure was the necessity of giving rights a foundation. We have gone some distance toward doing so by providing a framework within which rights claims can, in principle at least, be validated. Nihilism about rights has thereby been shown to be unjustified. The informed guesses and hunches of this chapter have been intended to suggest that scepticism about rights is also unjustified. We should not understate the obstacles to be overcome if we are to succeed in grounding a set of rights for any significant sector of our lives. But we also should not fear to make the attempt.

BIBLIOGRAPHY

ANDERSON, Alan R., 1962, 'Logic, Norms and Rules', *Ratio* 4, 1.

AUSTIN, John, 1885, *Lectures on Jurisprudence, or the Philosophy of Positive Law.* 5th Edn. Ed. Robert Campbell. London: John Murray.

BADHWAR, Neera K., 1985, 'Friendship, Justice, and Supererogation', *American Philosophical Quarterly* 22, 2.

BENTHAM, Jeremy, 1977, *A Comment on the Commentaries.* Ed. J. H. Burns and H. L. A. Hart. London: Athlone Press.

——1970(*a*), *Introduction to the Principles of Morals and Legislation.* Ed. J. H. Burns and H. L. A. Hart. London: Athlone Press.

——1970(*b*), *Of Laws in General.* Ed. H. L. A. Hart. London: Athlone Press.

——1952–4, *Jeremy Bentham's Economic Writings.* Three Vols. Ed. W. Stark. London: George Allen & Unwin.

——1843, *The Works of Jeremy Bentham.* Eleven Vols. Ed. John Bowring. Edinburgh: William Tait.

BERGER, Fred. R., 1984, *Happiness, Justice, and Freedom: The Moral and Political Philosophy of John Stuart Mill.* Berkeley: University of California Press.

BRANDT, Richard B., 1984, 'Utilitarianism and Moral Rights', *Canadian Journal of Philosophy* 14, 1.

BROWN, Stuart M., Jr., 1955, 'Inalienable Rights', *Philosophical Review* 64, 2.

BROWNLIE, Ian, 1981, *Basic Documents on Human Rights.* 2nd Edn. Oxford: Clarendon Press.

BUCHANAN, James M., 1975, *The Limits of Liberty: Between Anarchy and Leviathan.* Chicago and London: University of Chicago Press.

DICKENS, Bernard, 1979, *Guidelines on the Use of Human Subjects.* Toronto: Office of Research Administration, University of Toronto.

DONAGAN, Alan, 1977, *The Theory of Morality.* Chicago and London: University of Chicago Press.

DWORKIN, Ronald, 1977, *Taking Rights Seriously.* Cambridge, MA: Harvard University Press.

ELSTER, Jon, 1984, *Ulysses and the Sirens: Studies in Rationality and Irrationality.* Revised Edn. Cambridge: Cambridge University Press.

FEINBERG, Joel, 1980, *Rights, Justice, and the Bounds of Liberty: Essays in Social Philosophy.* Princeton: Princeton University Press.

FINNIS, John, 1983, *Fundamentals of Ethics.* Oxford: Clarendon Press.

——1980, *Natural Law and Natural Rights*. Oxford: Clarendon Press.

FITCH, Frederic B., 1967, 'A Revision of Hohfeld's Theory of Legal Concepts', *Logique et Analyse* 39–40.

FREY, R. G., 1984, 'Act-Utilitarianism, Consequentialism, and Moral Rights', in R. G. Frey, ed., *Utility and Rights* (Minneapolis: University of Minnesota Press).

GAUTHIER, David, 1986, *Morals by Agreement*. Oxford: Clarendon Press.

GIBBARD, Allan, 1984, 'Utilitarianism and Human Rights', *Social Philosophy & Policy* 1, 2.

GRAY, John, 1984, 'Indirect Utility and Fundamental Rights', *Social Philosophy & Policy* 1, 2.

——1983, *Mill on Liberty: A Defence*. London: Routledge & Kegan Paul.

GRIFFIN, James, 1984, 'Towards a Substantive Theory of Rights', in R. G. Frey, ed., *Utility and Rights* (Minneapolis: University of Minnesota Press).

GROTIUS, Hugo, 1925, *De Jure Belli ac Pacis Libri Tres*. Tr. Francis W. Kelsey. Oxford: Clarendon Press.

HACKER, P. M. S., 1973, 'Sanction Theories of Duty', in A. W. B. Simpson, ed., *Oxford Essays in Jurisprudence*. Second Series. Oxford: Clarendon Press.

HARE, R. M., 1981, *Moral Thinking: Its Levels, Method, and Point*. Oxford: Clarendon Press.

HARMAN, Gilbert, 1980, 'Moral Relativism as a Foundation for Natural Rights', *Journal of Libertarian Studies* 4, 4.

HARSANYI, John C., 1985, 'Rule Utilitarianism, Equality, and Justice', *Social Philosophy & Policy* 2, 2.

——1982, 'Morality and the Theory of Rational Behaviour', in Amartya Sen and Bernard Williams, eds., *Utilitarianism and Beyond*. Cambridge: Cambridge University Press.

HART, H. L. A., 1983, *Essays in Jurisprudence and Philosophy*. Oxford: Clarendon Press.

——1982, *Essays on Bentham: Studies in Jurisprudence and Political Theory*. Oxford: Clarendon Press.

——1961, *The Concept of Law*. Oxford: Clarendon Press.

——1958, 'Legal and Moral Obligation', in A. I. Melden, ed., *Essays in Moral Philosophy* (Seattle: University of Washington Press).

——1955, 'Are There Any Natural Rights?', *Philosophical Review* 64, 2.

HOBBES, Thomas, 1968, *Leviathan*. Ed. C. B. Macpherson. Harmondsworth: Penguin Books.

HOHFELD, Wesley Newcomb, 1919, *Fundamental Legal Conceptions As Applied in Judicial Reasoning*. Ed. Walter Wheeler Cook. New Haven and London: Yale University Press.

216   *Bibliography*

HUMANA, Charles, 1984, *World Human Rights Guide*. New York: Pica Press.

KANGER, Stig, and Helle KANGER 1966, 'Rights and Parliamentarism', *Theoria* 32, 2.

LEWIS, David K., 1969, *Convention: A Philosophical Study*. Cambridge, MA: Harvard University Press.

LINDAHL, Lars, 1977, *Position and Change: A Study in Law and Logic*. Dordrecht: D. Reidel Publishing Company.

LOCKE, John, 1954, *Essays on the Law of Nature*. Ed. W. von Leyden. Oxford: Clarendon Press.

——1894, *Essay Concerning Human Understanding*. Ed. A. C. Fraser. Oxford: Clarendon Press.

LYONS, David, 1982, 'Utility and Rights', in J. Roland Pennock and John W. Chapman, eds., *Nomos 24: Ethics, Economics and the Law* (New York and London: New York University Press).

——1978, 'Mill's Theory of Justice', in Alvin I. Goldman and Jaegwon Kim, eds., *Values and Morals* (Dordrecht: D. Reidel Publishing Company).

——1977(*a*), 'Human Rights and the General Welfare', *Philosophy & Public Affairs* 6, 2.

——1977(*b*), 'Principles, Positivism and Legal Theory', *Yale Law Journal* 87, 2.

——1976, 'Mill's Theory of Morality', *Nous* 10, 2.

——1970, 'The Correlativity of Rights and Duties', *Nous* 4, 1.

——1969, 'Rights, Claimants and Beneficiaries', *American Philosophical Quarterly* 6, 3.

McCLENNAN, Edward F., 1985, 'Prisoner's Dilemma and Resolute Choice', in Richmond Campbell and Lanning Sowden, eds., *Paradoxes of Rationality and Cooperation: Prisoner's Dilemma and Newcomb's Problem* (Vancouver: University of British Columbia Press).

MACCORMICK, Neil, 1982, *Legal Right and Social Democracy: Essays in Legal and Political Philosophy*. Oxford: Clarendon Press.

——1981, *H. L. A. Hart*. London: Edward Arnold.

——1978, *Legal Reasoning and Legal Theory*. Oxford: Clarendon Press.

——1977, 'Rights in Legislation', in P. M. S. Hacker and J. Raz, eds., *Law, Morality, and Society: Essays in Honour of H. L. A. Hart*. Oxford: Clarendon Press.

MACDONALD, Margaret, 1946–7, 'Natural Rights', *Proceedings of the Aristotelian Society* 47.

MACHAN, Tibor R., 1978, 'Against Nonlibertarian Natural Rights', *Journal of Libertarian Studies* 2, 3.

MACKIE, John, 1978, 'Can There Be a Right-Based Moral Theory?', *Midwest Studies in Philosophy* 3.

——1977, *Ethics: Inventing Right and Wrong*. Harmondsworth: Penguin Books.

MARTIN, Rex, 1985, *Rawls and Rights*. Lawrence, KS: University Press of Kansas.

MARTIN, Rex, and James W. NICKEL, 1980, 'Recent Work on the Concept of Rights', *American Philosophical Quarterly* 17, 3.

MILL, John Stuart, 1969, *Essays on Ethics, Religion and Society*. Collected Works 10. Ed. J. M. Robson. Toronto: University of Toronto Press.

MONTAGUE, Phillip, 1980, 'Two Concepts of Rights', *Philosophy & Public Affairs* 9, 4.

MORRIS, Christopher W., 1985, 'Natural Rights and Public Goods', in Thomas Attig *et al.*, eds., *The Restraint of Liberty* (Bowling Green Studies in Applied Philosophy 7).

NAGEL, Thomas, 1986, *The View from Nowhere*. New York and Oxford: Oxford University Press.

NARVESON, Jan, 1984, 'Contractarian Rights', in R. G. Frey, ed., *Utility and Rights* (Minneapolis: University of Minnesota Press).

NATIONAL COMMISSION FOR THE PROTECTION OF HUMAN SUBJECTS OF BIOMEDICAL AND BEHAVIORAL RESEARCH, 1978, *The Belmont Report: Ethical Principles and Guidelines for the Protection of Human Subjects of Research*. Washington: DHEW Publication No. (OS) 78-0012.

NOZICK, Robert, 1974, *Anarchy, State, and Utopia*. New York: Basic Books.

PARFIT, Derek, 1984, *Reasons and Persons*. Oxford: Clarendon Press.

PÖRN, Ingmar, 1970, *The Logic of Power*. New York: Barnes and Noble.

PUFENDORF, Samuel, 1934, *De Jure Naturae et Gentium Libri Octo*. Tr. C. H. and W. A. Oldfather. Oxford: Clarendon Press.

RAWLS, John, 1985, 'Justice As Fairness: Political not Metaphysical', *Philosophy & Public Affairs* 14, 3.

——1971, *A Theory of Justice*. Cambridge, MA: Harvard University Press.

RAZ, Joseph, 1984(*a*), 'On the Nature of Rights', *Mind* 93, 370.

——1984(*b*), 'Legal Rights', *Oxford Journal of Legal Studies* 4, 1.

——1980, *The Concept of a Legal System: An Introduction to the Theory of Legal System*. 2nd Edn. Oxford: Clarendon Press.

——1979, *The Authority of Law: Essays on Law and Morality*. Oxford: Clarendon Press.

——1975, *Practical Reason and Norms*. London: Hutchinson.

ROBINSON, R. E., *et al.*, 1983, 'The Logic of Rights', *University of Toronto Law Journal* 33.

ROSS, Alf, 1968, *Directives and Norms*. London: Routledge & Kegan Paul.

SCHEFFLER, Samuel, 1982, *The Rejection of Consequentialism.* Oxford: Clarendon Press.

SIDGWICK, Henry, 1962, *The Methods of Ethics.* 7th Edn. London: Macmillan and Company.

SIEGHART, Paul, 1985, *The Lawful Rights of Mankind.* Oxford: Oxford University Press.

——1983, *the International Law of Human Rights.* Oxford: Clarendon Press.

SLOTE, Michael, 1985, *Common-sense Morality and Consequentialism.* London: Routledge & Kegan Paul.

SOLES, David E., 1985, 'Sentience and Moral Standing', *Dialogue* 24, 4.

SOUSA, Ronald de, 1985, 'Arguments from Nature', in David Copp and David Zimmerman, eds., *Morality, Reason and Truth: New Essays on the Foundations of Ethics* (Totowa, NJ: Rowman & Allanheld).

STOLJAR, Samuel J., 1984, *An Analysis of Rights.* London and Basingstoke: The Macmillan Press.

SUMNER, L. W., 1985, 'Moral Theory and Moral Standing: A Reply to Woods and Soles', *Dialogue* 24, 4.

——1981, *Abortion and Moral Theory.* Princeton: Princeton University Press.

TOOLEY, Michael, 1972, 'Abortion and Infanticide', *Philosophy & Public Affairs* 2, 1.

TUCK, Richard, 1979, *Natural Rights Theories: Their Origin and Development.* Cambridge: Cambridge University Press.

WALDRON, Jeremy, 1981, 'A Right to do Wrong', *Ethics* 92, 1.

WELLMAN, Carl, 1985, *A Theory of Rights: Persons Under Laws, Institutions, and Morals.* Totowa, NJ: Rowman & Allanheld.

WHITE, Alan R., 1984, *Rights.* Oxford: Clarendon Press.

WILLIAMS, Bernard, 1973, 'A Critique of Utilitarianism', in J. J. C. Smart and Bernard Williams, *Utilitarianism: For and Against* (Cambridge: Cambridge University Press).

WORKING GROUP ON HUMAN EXPERIMENTATION, 1978, *Ethical Considerations in Research Involving Human Subjects.* Ottawa: Medical Research Council Report No. 6.

INDEX